# 和全球@做生意

# 必備商用英文

# E-mail

## An Effective Guide to Business Writing

U0131504

發 行 人　鄭俊琪

編　　著　希伯崙編輯團隊

總 編 輯　陳豫弘

責任編輯　林詩嘉

英文編輯　Zach Mortimer・Ana Avvakumova・Dilan Schulte

英文錄音　Mike Tennant・Ashley Smith・Meagunn Hart・Doug Nienhuis

美術編輯　黃薪宸

封面設計　黃薪宸

點讀製作　翁稚緹

出版發行　希伯崙股份有限公司

　　　　　105 台北市松山區八德路三段 32 號 12 樓

　　　　　電話：(02) 2578-7838

　　　　　傳真：(02) 2578-5800

　　　　　電子郵件：service@liveabc.com

法律顧問　朋博法律事務所

印　　刷　禹利電子分色有限公司

出版日期　2021 年 5 月　初版一刷

# 和全球做生意
# 必備商用英文
# E-mail

## An Effective Guide to Business Writing

英語數位學習第一品牌

# CONTENTS

如何使用本書 **8**

社交公關篇

## 如何使用本書

# 精通英文商業寫作，強化商務英語溝通力！

遇到講求特定格式、修辭、句構和習慣用語的英文商務書信，總會有「寫」到用時方恨少的無力感嗎？別擔心，本書能幫你擺脫總是寫出職場囧英文的行列。

本書分為【寫作入門篇】、【人事行政篇】、【工作業務篇】、【社交公關篇】及【貿易往來篇】5 大主題。【寫作入門篇】帶你熟悉商務 e-mail 的結構、稱謂／敬語、內文格式、起頭句和結尾句，讓你打下商業寫作的基礎。【人事行政篇】、【工作業務篇】、【社交公關篇】及【貿易往來篇】則依照使用情境，各收錄 9 種實用商務書信，涵蓋共 36 個單元。每個單元精選 1-2 封範例信，並搭配主題詞彙、重點句型、寫作要點和翻譯練習等學習元素，以深入淺出的方式，有效強化你的英文商業書信寫作技巧。

**暖身活動** 進入課程內容前，先透過翻譯練習和關鍵字詞對各篇主題有初步的認識。

掃描 QR Code 即可聆聽全篇內容。

每篇規劃 4 個翻譯例句，培養基本概念。

每篇精選 8 個主題詞彙，打下學習基礎。

每篇收錄 **4** 個關鍵句型，並透過例句，正確理解句型的用法。

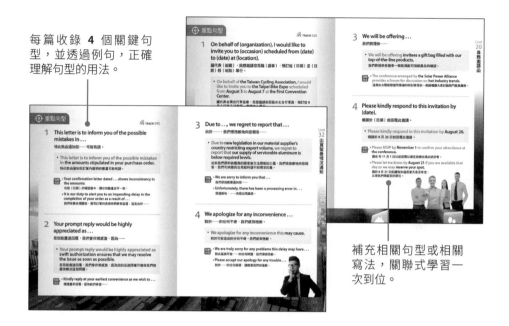

補充相關句型或相關寫法，關聯式學習一次到位。

每篇收錄 **1-2** 個應用範例，讓你通透各主題信件的寫作模式。

拉框補充說明，加深學習印象。

中英對照，提升理解信件內容的效率。

彙整實用單字和片語，增加字彙量。

延伸學習助你理解各類書信的寫作重點。

# 點讀筆功能介紹＋MP3 / 點讀音檔下載

## 認識點讀筆

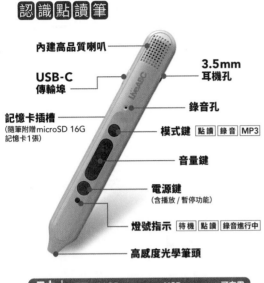

- 內建高品質喇叭
- USB-C 傳輸埠
- 3.5mm 耳機孔
- 記憶卡插槽（隨筆附贈microSD 16G 記憶卡1張）
- 錄音孔
- 模式鍵　點讀　錄音　MP3
- 音量鍵
- 電源鍵（含播放 / 暫停功能）
- 燈號指示　待機　點讀　錄音進行中
- 高感度光學筆頭

三大特色｜16GB 記憶卡｜USB Type-C｜可充電鋰電池

## 四大功能

- ◆ 點讀發音
- ◆ 錄音發音
- ◆ MP3 播放
- ◆ 英漢字典

| | | |
|---|---|---|
| 高科技光學點讀筆頭 | 尺寸 | 151 x 20 x19 mm |
| 內建高品質喇叭 | 重量 | 36±2g（內含鋰電池） |
| 支援USB檔案傳輸 | 記憶體 | 含 16GB microSD 記憶卡 |
| | 電源 | 鋰電池（500mAH） |
| 4 in one 點讀/錄音 MP3/字典 四機一體 | 配件 | USB 傳輸線 (Type-C Cable)、使用說明書、錄音卡 / 音樂卡 / 字典卡、microSD 記憶卡（已安裝） |

## MP3 音檔下載及安裝步驟

**Step 1**
在 LiveABC 首頁上方的「叢書館」點擊進入。點選「上班族英語」。找到您要下載的書籍後，點擊進入內容介紹網頁。

**Step 2**
點選內容介紹裡的「MP3 音檔下載」，進行下載該書的 MP3 壓縮檔。將下載好的壓縮檔解壓縮後，會得到一個資料夾，裡面即有本書的 MP3 音檔。

**Step 3**
將下載好的 MP3 音檔存放於電腦或手機裡，即可直接播放本書內容，加強您的聽力學習。

## 點讀筆音檔下載及安裝步驟

**Step 1**
在 LiveABC 首頁上方的「叢書館」點擊進入。點選「上班族英語」。找到您要下載的書籍後，點擊進入內容介紹網頁。

**Step 2**
點選內容介紹裡的「點讀筆音檔下載」，找到您需要的點讀音檔後點一下，進行下載點讀筆壓縮檔。將下載好的壓縮檔解壓縮後，即能得到本書的點讀筆音檔。

**Step 3**

用 USB 傳輸線連結電腦和點讀筆，會出現「點讀筆」資料夾，點擊兩下進入「BOOK」資料夾。

**Step 4**

將解壓縮後所得到的點讀筆音檔按右鍵複製，然後在上一步驟的「BOOK」資料夾裡按右鍵貼上，即可完成點讀筆音檔的安裝。

# 開始使用點讀筆

適用
LiveABC
點讀筆

馬上就可套用，商用文書必備
教你寫出專業又不失禮貌的英文 E-mail

和全球做生意
必備**商用英文**
**E-mail**
An Effective Guide to Business Writing

4 大類別、65 篇實用範例，大方幫你搞定你所需要的
人事行政、工作業務、社交公關、貿易往來

LiveABC

## Step 1

1. 將 LiveABC 光學筆頭指向本書封面圖示。
2. 聽到「Here We Go!」語音後即完成連結。

Step 2　開始使用書中的點讀功能。

點 🎧 TRACK 007 圖示，即可聆聽本頁應用範例與教學內容的發音。

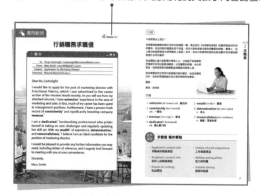

★ 每本書可點讀的內容依該書編輯規劃為準。

# 搭配功能卡片使用

RECORD & PLAY
錄音卡

MUSIC PLAYER
音樂卡

## 錄音功能　請搭配錄音卡使用

模式切換：　點選 RECORD & PLAY 錄音卡 ，聽到「Recording Mode」表示已切換至錄音模式。

開始錄音：　點選 ⦿，聽到「Start Recording」開始錄音。

停止錄音：　點選 ⦿⦿，聽到「Stop Recording」停止錄音。

播放錄音：　點選 ▶，播放最近一次之錄音。

刪除錄音：　刪除最近一次錄音內容，請點選 🗑 。
　　　　　　（錄音檔存於資料夾「\recording\meeting\」）

## MP3 功能　請搭配音樂卡使用

模式切換：　點選 MUSIC PLAYER 音樂卡 ，並聽到「MP3 Mode」表示已切換至 MP3 模式。

開始播放：　點選 ▶，開始播放 MP3 音檔。

新增／刪除：　請至點讀筆資料夾位置「\music\」新增、刪除 MP3 音檔。

DICTIONARY
KEYPAD
英漢字典卡

## 英漢字典功能　請搭配字典功能版使用

模式切換：　點選 Dictionary ON ，聽到「Dictionary on」表示已切換至字典模式。

單字查詢：　依序點選單字拼字，完成後按 Enter ，即朗讀字彙的英語發音和中文語意。

關閉功能：　使用完畢點選 Dictionary OFF ，即可回到點讀模式。

更多點讀筆使用說明請掃描 QR code

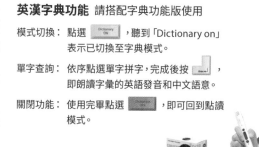

# Business Writing 101
## 寫作入門篇

主題詞彙

圖解
E-mail 結構

主旨

稱謂

內文格式

結尾敬語

附加檔
常見句型

起頭
好用句型

結尾
好用句型

# Business E-mail 101

## 商務 E-mail 基本知識

## 課前寫作練習

商務 E-mail 怎麼寫？
請參考主題詞彙、中譯及括號內的英文提示，將下列句子翻譯成英文。

❶ 這封郵件旨在通知您敝司的人事異動。

寫 _____

_____

主題詞彙

- **addressee/recipient**
  [ˌædrɛˋsi] [rɪˋsɪpɪənt] 收件人
- **attach** [əˋtætʃ]
  附件

- **blind carbon copy**
  [ˋkɑrbən] 密件副本
- **carbon copy**
  副本

❷ 謹代表萊夫供應商（Life Supplies）特此來信。

（寫）

_____

_____

❸ 請見附加檔我們新的 HandsOn 產品簡介。

（寫）

_____

_____

❹ 如果您還有其他問題或疑慮，請盡管和我聯絡。

（寫）

_____

_____

參考答案請見 p. 19

- **complimentary close**
  [ˌkɑmpləˋmɛntəri] 結尾敬語
- **look forward to**
  期盼

- **salutation** [ˌsæljəˋteʃən]
  稱謂
- **subject** [ˋsʌbˌdʒɛkt]
  主旨

# 圖解 E-mail 結構

**❶** To:  Kevin Tempo <ktempo@topdesign.com>

**❷** CC:  Norman Cook <ncook@elite.com >

BCC:

**❸** Subject:  Away on Business

**❹** Attached:  contact list.docx

**❺** Dear Mr. Tempo:

I'm away on business from February 6 to 8 and won't be able to respond to you until I get back. Please feel free to leave me a message, and I promise I'll get back to you as soon as I am able.

**❻** If you are calling about a matter of importance, please contact Lindy Layton at 0208-208-0408 or lindy@dubbeats.com. She will be covering for me while I'm away.

**❼** Sincerely,

Norman Cook

## 基本結構

**❶ Addressee and E-mail Address** 收件人和電子郵件地址

**❷ CC (carbon copy)** 副本
**BCC (blind carbon copy)** 密件副本
carbon 原指「複寫紙」，在 e-mail 中，carbon copy 的功能是將信件「副本抄送」給其他人。blind carbon copy 則讓信件以隱藏收件人地址的方式副本抄送。

**❸ Subject** 主旨
點出信件主要內容。收件人常藉由主旨決定是否閱讀該信，或判斷其重要性，故主旨寫法應簡短具體。

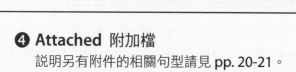

## 中譯

收件者：凱文‧田波 <ktempo@topdesign.com>
副本：諾門‧庫克 <ncook @elite.com>
密件副本：
主旨：出差中
附件：**contact list.docx**

田波先生您好：

**2 月 6 日**至 **8 日**我正在出差，直到回來前都無法回覆您。
歡迎留言給我，我保證只要我有空時就會答覆您。

如果您有要事須來電，請撥 **0208-208-0408** 或寄信到
**lindy@dubbeats.com** 聯絡琳蒂‧雷頓。我不在時她會代
為處理我的工作。

謹啟，

諾門‧庫克

---

❹ **Attached** 附加檔
　說明另有附件的相關句型請見 **pp. 20-21**。

❺ **Salutation** 稱謂
　可視與對方的關係來決定，或直接以名字稱呼，有時也會省略稱謂或名字而直
　接進入 e-mail 內文（詳見 **p. 18**）。

❻ **Body** 內文
　e-mail 的主要訊息，格式與寫法請見 **p. 18**。

❼ **Complimentary Close** 結尾敬語
　對象不同也會有不同的結尾敬語（詳見 **p. 19**）。

| 稱呼寫法 | 說明 |
|---|---|
| Dear Mr. _____ | • 商務往來或較正式信函的稱呼盡可能寫出稱謂，稱謂語之後一般接逗號或冒號。 |
| Dear Mrs. _____ | • Mr. 為男性稱謂。 |
| | • Mrs. 為知悉對方已婚時適用。 |
| Dear Miss _____ | • Miss 為知悉對方未婚時適用。 |
| | • Ms. 是已婚或未婚女性皆可使用。 |
| Dear Ms. _____ | |
| Dear John & Mary<br>Dear all | 同時寫信給兩個人可直接寫出名字，同時寄信給多人時則可用 Dear all。 |
| Dear Sir or Madam | 無法得知收件人姓名時可使用。 |

## 內文格式與寫法

| 格式 | 說明 |
|---|---|
| fully blocked layout<br>齊頭式 | 內文全部靠左對齊的書寫格式。 |
| itemized style<br>條列式 | 如有若干建議或事項要闡述，可用條例式的寫法，會更清楚閱讀。 |
| simplification<br>簡化 | 因 e-mail 多為非正式文件，對格式和寫法的要求較不嚴格，常可簡化句子或使用簡寫。<br>常用簡寫範例<br>• ASAP = as soon as possible 盡快<br>• FYI = for your information 供參<br>• BTW = by the way 順帶一提 |

## 結尾敬語

依寫信對象的不同，結尾敬語也會不同。結尾敬語大多沒有固定的翻法，
一般常譯作謹致、敬啟、謹啟或謹上。以下列舉一些常見的寫法：

formal

Sincerely, 謹啟
Sincerely yours, 敬啟
Yours sincerely, 敬啟

⇨ 正式用法。在商業書信中，若和對方沒有個別交情，只是商務往來的關係，用 Sincerely 最妥當。

Best regards, 謹致
Regards, 謹致
Best wishes, 最深的祝福

⇨ 可用於有建立關係的客戶、長官或同事。

Truly yours, 真誠地
Yours truly, 真誠地
Faithfully yours, 真誠地
Yours faithfully, 真誠地
Cordially yours, 謹上
Yours cordially, 謹上
Best, 祝一切安好
Yours, 謹上
Cheers, 祝愉悅
Take care, 保重

⇨ 非正式用法，可用於親友或熟稔的同事間。

informal

**參考答案**

1. This e-mail is to inform you of the personnel changes in our company.

2. I'm writing this letter on behalf of Life Supplies.

3. Please see/find the attached overview of our new HandsOn product.

4. If you have any more questions or concerns, please don't hesitate to contact me.

## 附加檔

在 e-mail 內文中可提醒收件者另有附件，電子郵件的附件一般都用 attach，
一般郵寄信件中所附的資料則常用 enclose。說明 e-mail 附件的常見句型有：

## 1 Please see/find the attached . . .

請見附加檔的……

- **Please find the attached overview of our new HandsOn product.**
  請見附加檔我們新的 HandsOn 產品簡介。

## 2 . . . are attached.

……已附上。

- **Quotations for models A-215, A-228, and D-448 are attached.**
  機型 A-215、A-228 和 D-448 的報價已附上。

## 3 I've attached . . .

我已附上……

- **I've attached some of the pictures we took during our vacation.**
  我已附上一些我們度假時所拍的照片。

## 4 Attached is . . . for your perusal.

附上……供您詳閱。

- **Attached is the trip report for your perusal.**
  附上這次的出差報告供您詳閱。

## 5 Per your request, attached is . . .

如您所要求，附件是……

- **Per your request, attached is our latest catalogue.**
  如您所要求，附件是我們最新的產品型錄。

## 6 I have also included . . .

我也附上……

- **I have also included our current price list.**
  我也附上我們現行的價目表。

## 7 Attached please find . . .

附件可看到……

- **Attached please find an itemized estimate of all materials and labor plus information on installation services.**
  附件可看到全部原料和勞力的預估明細外加裝修服務的資訊。

## 1 This e-mail is to inform you of . . .

這封郵件旨在通知您……

> • **This e-mail is to inform you of the personnel changes in our company.**
> 這封郵件旨在通知您敝司的人事異動。

 相關句型
• **I am writing to inquire . . .**
來信是要詢問……

• **I am writing to ask if . . .**
來信詢問是否……

## 2 The purpose of this letter is to . . .

此信旨在……

> • **The purpose of this letter is to ascertain if you are capable of supplying your latest model.**
> 此信旨在確認您是否能供應貴司的最新機種。

 相關句型
• **The reason for this letter is to learn . . .**
來信的原因是要知道……

• **I am writing to learn . . .**
來信是要知道……

## 3 I'm writing this letter on behalf of (company/organization).

謹代表（公司／機構）特此來信。

- **I'm writing this letter on behalf of Life Supplies.**
  謹代表萊夫供應商特此來信。

- **On behalf of (company/organization), I am writing to inform you that . . .**
  謹代表（公司／機構），來信告知您……

- **We at (company/organization) would like to . . .**
  （公司／機構）的全體員工想……

## 4 This is in response to . . .

此信旨在回應……

- **This is in response to your recent advertisement for a senior engineer position.**
  此信旨在回應貴司最近的資深工程師一職的廣告。

- **In response to your letter of (date), we are pleased to . . .**
  為回應您在（日期）的來信，
  我們很高興能……

23

## 5 Thank you very much for ...

非常感謝您……

- **Thank you very much for your recent order for 200 of our bed coverings.**
  非常感謝您最近訂購兩百件我們的床罩商品。

 相關句型
- **We were pleased to receive your letter on (date) for ...**
  我們很高興在（日期）收到您……的來信。
- **Thank you for your letter on (date) ...**
  感謝您於（日期）來信……

---

## 6 Thank you for giving me the opportunity of ...

感謝您給我機會……

- **Thank you for giving me the opportunity of meeting with you on Wednesday.**
  感謝您給我機會於週三和您會面。

 相關寫法
- **It was a great pleasure to meet you last week and to discuss your advertising campaign for the Seek series.**
  很高興上週能和您見面並討論貴司「尋覓」系列產品的廣宣活動。

# 7 With regard to . . .

關於……

- **With regard to your delivery scheduled for next Wednesday, I regret to inform you that the shipping will be delayed.**

  關於您預計於下週三運送的貨品，很抱歉要告知您運貨將會延遲。

- **In light of the many requests for extended operating hours, we will initiate new service hours beginning the first of next month.**

  有鑒於很多關於延長營運時間的需求，我們將自下個月 1 日起實施新的服務時段。

---

# 8 Per your request, . . .

依據您的要求，……

- **Per your request, attached are the materials about cyber-commuting.**

  依據您的要求，附上電子通勤的相關資料。

- **As requested, I have sent out the new samples by courier this morning.**

  如您所要求，我今天早上已請快遞送出新的樣品。

# 1 I look forward to . . .

期盼……

> • **I look forward to hearing from you soon.**
> 期盼盡快聽到您的回應。

**look forward to + N./V-ing** 可說是信件結尾最常用的基本句型，來看看以下其他的例子吧：

**I look forward to**
期盼

- **your response.**
  收到您的回覆。

- **receiving your counsel.**
  得到您的建議。

- **meeting with you soon.**
  盡快與您會面。

- **speaking with you soon.**
  盡快與您談話。

- **hearing what you think of this.**
  聽到您對這件事情的想法。

- **discussing this at your earliest convenience.**
  在您有空時盡快與您討論此事宜。

## 2 If you have . . ., please don't hesitate to . . .

如果您有……，請不吝……

- If you have **any more questions or concerns**, please don't hesitate to **contact me**.

  如果您還有其他問題或疑慮，請不吝和我聯絡。

- Should you require . . ., feel free to . . .

  若您需要……，歡迎……

- Contact me at any time should you have . . .

  如果您有……，請隨時與我聯絡。

- Please let me know if I can do anything else to . . .

  如果有其他我能……的地方，請讓我知道。

---

## 3 We are privileged to . . .

我們深感榮幸能……

- We are privileged to be your supplier and look forward to providing continued assistance and service.

  我們深感榮幸能擔任你們的供應商，並期盼能繼續提供協助和服務。

- We have the honor of . . .

  我們很榮幸能……

- We are grateful for . . .

  我們很感謝能……

- We are truly honored to . . .

  我們真的很榮幸能……

## 4 Please let us know at your earliest convenience whether . . .

請盡早讓我們知道您是否……

- **Please let us know at your earliest convenience whether you will be attending this workshop.**
  請盡早讓我們知道您是否會參加這場專題討論會。

 相關
句型
- **Your prompt attention to . . . will be highly appreciated.**
  我們會非常感謝您即刻處理……
- **Please contact us when you've had time to . . .**
  麻煩您有空……時與我聯繫。

---

## 5 We would appreciate it if you could . . .

如果您能……，我們將不勝感激。

- **We would appreciate it if you could let us know when you sent the products.**
  如果您能讓我們知道您何時寄出產品，我們將不勝感激。

 相關
句型
- **Is it possible for you to . . .?**
  能否請您……？
- **Please advise us as to . . .**
  請告知關於……

## 6 We hope that our . . . will lead to . . .

期盼我們的……能導向……

- **We hope that our prompt and efficient handling of your order will lead to your becoming a regular customer.**

  期盼我們迅速且有效率地處理您的訂單能讓您成為我們的老主顧。

 **相關寫法** • **We hope that we will have the pleasure of receiving further orders from you.**

  我們希望能有這個榮幸再收到更多貴司的訂單。

---

## 7 We very much appreciate your . . . and wish you . . .

我們非常感謝您……，也祝福您……

- **We very much appreciate your patronage and wish you every personal and professional success in the future.**

  我們非常感謝您的惠顧，也祝福您未來在個人和事業上取得成功。

 **相關寫法** • **We wish you all the best in your career and future endeavors.**

  祝福您在職涯以及未來的努力上一切順利。

# Personnel Administration
## 人事行政篇

# Unit 2

# Application Letters
## 求職信

## 課前寫作練習

求職信怎麼寫？
請參考主題詞彙、中譯及括號內的英文提示，將下列句子翻譯成英文。

❶ 我想要應徵快樂大賣場（Happy Hypermarket）的採購經理一職，我在 JobSearch.com 網站上看到刊登的廣告。

寫

主題
詞彙

- **advertise** [`ædvɚ͵taɪz]
  登廣告
- **apply for** [ə`plaɪ]
  申請
- **candidate** [`kændə͵det]
  應試者
- **post/position** [pə`zɪʃən]
  職務；職位

❷ 從我的履歷表您可知道，我有公關方面的經驗。

(寫)

_____

_____

❸ 我是個具前瞻性思考力的專業人士，以能做出實質貢獻感到驕傲。

(寫)

_____

_____

❹ 以我的財務規劃技巧，我相信我是投資顧問這項職務的理想人選。

(寫)

_____

_____

參考答案請見 p. 39

- **professional** [prəˋfɛʃən!]
  專業人士
- **reference** [ˋrɛfərəns]
  推薦信

- **résumé** [ˋrɛzə͵me]
  履歷
- **track record** [træk] [ˋrɛkəd]
  過往紀錄

**1** **I would like to apply for the post of (title) with (company), which I saw advertised in/on . . .**

我想要應徵（公司名）的（職稱）一職，我在……看到刊登的廣告。

- I would like to apply for the post of **project manager** with **Delta Home Entertainment Systems,** which I saw advertised on **NeedJob.com.**

  我想要應徵戴達家庭娛樂系統公司的專案經理一職，我在 NeedJob 網站上看到刊登的廣告。

 • In response to your advertisement on **Job411.com** for a **management trainee** position, I am attaching my résumé.

  為回應貴司在 Job411.com 網站刊登儲備幹部一職的徵才廣告，隨函附上我的履歷。

**2** **As you will see from my résumé, I have experience in the area of . . .**

從我的履歷表您可知道，我有……方面的經驗。

- As you will see from my résumé, I have experience in the area of **event planning.**

  從我的履歷表您可知道，我有活動企畫方面的經驗。

 • As you will see from my résumé, I have **seven** years' experience in **public relations.**

  從我的履歷您可知道，我有七年的公關經驗。

# 3 I am a(n) . . . professional who prides himself/ herself in . . .

我是個……的專業人士，以能……感到驕傲。

- I am a **detail-oriented** professional who prides himself in **maintaining quality at work.**

  我是個細節取向的專業人士，以能維持工作品質感到驕傲。

 • I am certain that I would make a valuable addition to **BeInvention Ltd.** because of **my in-depth knowledge of the industry.**

  有鑒於自己對這個產業的深入了解，我相信我會是 BeInvention 有限公司不可多得的人才。

---

# 4 With my . . . , I believe I am an ideal candidate for the position of (title).

以我的……，我相信我是（職稱）這項職務的理想人選。

- With my **interpersonal skills,** I believe I am an ideal candidate for the position of **PR manager.**

  以我的人際關係技巧，我相信我是公關經理這項職務的理想人選。

 • I look forward to meeting with you to **discuss this possibility.**
  期盼能與您會面討論這個可能性。

- I would be pleased to provide **any further information you may need** and hope to **hear from you soon.**
  我很樂意提供任何您可能需要的額外資訊並期盼盡快聽到您的消息。

# 行銷職務求職信

To: Tanya Cartwright <tcartwright@functionalfabrics.com>
From: Mary Smith <msmith@job411.com>
Subject: Application for Marketing Director
Attached: Résumé-MarySmith.docx

Dear Ms. Cartwright:

I would like to apply for the post of marketing director with Functional Fabrics, which I saw advertised in the career section of the *Houston Herald* recently. As you will see from my attached résumé, I have **extensive**[1] experience in the area of marketing and sales. In fact, much of my career has been spent in management positions. Furthermore, I have a proven track record of **consistently**[2] and significantly boosting company **revenue**.[3]

I am a **dedicated**,[4] hardworking professional who prides herself in taking on new challenges and regularly updating her skill set. With my **wealth**[5] of experience, **determination**,[6] and **resourcefulness**,[7] I believe I am an ideal candidate for the position of marketing director.

I would be pleased to provide any further information you may need, including letters of reference, and I eagerly look forward to meeting with you at your convenience.

Sincerely,

Mary Smith

## 中譯

卡特萊特女士您好：

我想要應徵機能紡織公司的行銷總監一職，最近我在《休斯頓先鋒報》的職涯版看到刊登的廣告。從我所附的履歷表您可知道，我在行銷和銷售方面有很豐富的經驗。事實上，我大部分的職涯都是在管理職位上度過。此外，我有可供證明的過往紀錄來顯示我持續並顯著提升公司的營收。

我是個盡心盡力且勤奮的專業人士，以能接下新挑戰和定期提升自我技能感到驕傲。以我豐富的經驗、決心和智謀，我相信我是行銷總監這項職務的理想人選。

我很樂意提供任何您可能需要的額外資訊，包括推薦信在內，我也非常期盼能在您方便的時間與您會晤。

謹啟，

瑪莉‧史密斯

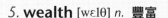

## 字彙

1. **extensive** [ɪkˋstɛnsɪv] *adj.* 廣泛的

2. **consistently** [kənˋsɪstəntli] *adv.* 一貫地

3. **revenue** [ˋrɛvə͵nju] *n.* 營收

4. **dedicated** [ˋdɛdə͵ ketəd] *adj.* 盡心盡力的

5. **wealth** [wɛlθ] *n.* 豐富

6. **determination** [dɪ͵tɝməˋneʃən] *n.* 決心

7. **resourcefulness** [rɪˋsorsfəlnəs] *n.* 機敏；足智多謀

## 延伸學習　求職信 寫作要點

- **Applicant's contact info**
  求職者的聯絡資訊

- **Recipient's contact info**
  收件人的聯絡資訊

- **Reason for writing**
  致函原因

- **Outline of work experience**
  工作經歷概述

- **Abilities and qualities**
  能力與特質

- **Suitable closing**
  適當的結語

# 業務職缺求職信

| To: | Tammy Sawyer <tsawyer@get-go.com> |
| From: | Robert Ellis <robert.ellis@gmail.com> |
| Subject: | Application for International Sales Position |
| Attached: | Résumé-RobertEllis.docx |

Dear Ms. Sawyer,

In response to your advertisement on Jobs4U.com for an international sales position, I am attaching my résumé.

In my most recent position at SouthTech Corp., I rose from a sales assistant to a sales manager. As you will see from my résumé, I **spearheaded**[1] SouthTech Corp.'s social network marketing efforts, **overseeing**[2] the establishment of both a Facebook and a Twitter **presence**.[3]

I understand that Get-Go Industries is looking to expand its business in Asia, particularly in China. Having graduated in international business with a **concentration**[4] on Asia, I spent a year studying in China and am **fluent**[5] in Mandarin Chinese. With this in mind, I believe I would be the perfect person to **facilitate**[6] your **push**[7] into the China market.

> with sth in mind 指「考慮到……；有鑒於……」。

I look forward to meeting with you to discuss this possibility. If you need me to provide any additional information or references, don't hesitate to contact me.

Sincerely,

Robert Ellis

## 中譯

索爾女士您好，

為回應貴司在 **Jobs4U.com** 網站刊登國際業務一職的徵才廣告，隨函附上我的履歷。

在我上一個任職的南方科技公司裡，我從業務助理晉升為業務經理。從我的履歷您可知道，我帶領南方科技公司去推動社群網絡行銷，監督公司在臉書和推特建立能見度。

我明白開端工業期盼將業務拓展至亞洲，尤其是中國地區。自國貿系畢業，主修亞洲區域的我曾在中國攻讀一年，華語非常流利。有鑒於此，我相信我會是幫助貴司推進中國市場的絕佳人選。

我期盼能與您會面討論這個可能性。如果您需要我提供任何進一步的資訊或推薦信，請隨時與我聯繫。

謹啟，

羅伯特·艾利斯

## 字彙

1. **spearhead** [ˋspɪrˌhɛd]
   v. 充當……的先鋒

2. **oversee** [ˌovɚˋsi] v. 監督

3. **presence** [ˋprɛzn̩s] n. 存在

4. **concentration** [ˌkɑnsənˋtreʃən]
   n. 專注

5. **fluent** [ˋfluənt]
   adj. （語言）流利的

6. **facilitate** [fəˋsɪləˌtet] v. 促進

7. **push** [puʃ] n. 推進；奮進

---

**參考答案**

1. I would like to apply for the post of procurement manager with Happy Hypermarket, which I saw advertised on JobSearch.com.

2. As you will see from my résumé, I have experience in the area of public relations.

3. I am a forward-thinking professional who prides himself/herself in making substantial contributions.

4. With my financial planning skills, I believe I am an ideal candidate for the position of investment consultant.

# Unit 3 Recommendation Letters

## 推薦信

---

## 課前寫作練習

推薦信怎麼寫？
請參考主題詞彙、中譯及括號內的英文提示，將下列句子翻譯成英文。

❶ 我很榮幸能與席薇亞・西姆斯（Sylvia Sims）在羅斯頓皇家公司（Rolston Royal）共事三年。

寫 _____

_____

---

主題詞彙

- **application** [ˌæpləˋkeʃən]
  應徵；申請
- **asset** [ˋæˌsɛt]
  有用之才

- **candidacy** [ˋkændədəsi]
  候選資格
- **commitment** [kəˋmɪtmənt]
  奉獻精神

❷ 馬克‧惠勒（Mark Wheeler）在大眾圖書（Books for All）最初是行銷助理，並逐步晉升成為行銷經理。

Ⓦ _____

_____

❸ 由於白琳達‧卡爾森（Belinda Carson）的採購專長，她得以改善產品採購的效率。

Ⓦ _____

_____

❹ 我有信心布洛克‧海斯丁斯（Brock Hastings）在 EVO 公司（EVO Corporation）銷售團隊中，將是個寶貴人才。

Ⓦ _____

_____

參考答案請見 p. 47

- **contribution** [ˌkɑntrəˈbjuʃən]
  貢獻
- **dedication** [ˌdɛdəˈkeʃən]
  專心致力

- **endorse** [ɪnˈdɔrs]
  背書
- **recommend** [ˌrɛkəˈmɛnd]
  推薦

TRACK 009

**1** **I have had the pleasure of working alongside (name) for (period of time) at (company).**

我很榮幸能與（人名）在（公司）共事（一段時間）。

- **I have had the pleasure of working alongside Daniel Greer for 10 years at Bell Tower.**

  我很榮幸能與丹尼爾 · 格里爾在鐘塔公司共事十年。

 • **The purpose of this mail is to endorse (name)'s application for a position at your company.**

此郵件旨在背書（人名）於貴司的職缺應徵。

---

**2** **(Name) began as (title) at (company) and worked his/her way up to become (role).**

（人名）在（公司）最初是（職稱），並逐步晉升成為（職位）。

- **Matthew Wallace began as a junior purchaser at Palmer Inc. and worked his way up to become the head of Purchasing.**

  馬修 · 瓦里斯在帕爾瑪公司最初是初級採購員，並逐步晉升成為採購主管。

 • **(Name) became (role) after spending (time) climbing the career ladder at (company).**

花了（時間）在（公司）一路晉升後，（人名）當上（職位）。

- **(Name) was employed as (title) at (company) from (year) to (year).**

（人名）從（年分）至（年分）受雇於（公司）擔任（職位）。

## 3 With his/her expertise in . . . , (name) has been able to . . .

由於（人名）的……專長，他／她得以……

- With her expertise in **industrial management, Joanna McBride** has been able to **improve efficiency at the factory by 15 percent.**

  由於喬安娜‧麥克布萊德的工業管理專長，她得以改善廠房的效率達15%。

 • As an invaluable member of our team, (name) helped enable . . .

（人名）是這個團隊的寶貴成員，促成……

- In his/her (number) years at the company, (name) made a substantial contribution to . . .

  在他／她任職的（年數）裡，（人名）對……貢獻卓著。

---

## 4 I am confident that (name) will be an asset to . . .

我有信心（人名）在……中，將是個寶貴人才。

- I am confident that **Melvin Summers** will be an asset to **the engineering team at Smith Tech.**

  我有信心馬文‧桑莫斯在史密斯科技的工程團隊中，將是個寶貴人才。

 • I believe that (name) will do as sterling a job for (new company) as he/she did for (old company).

我相信（人名）將在（新公司）表現得與他／她在（舊公司）一樣優秀。

- I wholeheartedly support (name)'s candidacy at (new company).

  我真心支持（人名）在（新公司）的求職申請。

# 前上司推薦信

---

To: Teresa Curry <tcurry@vexinternational.com>
From: Josh Floyd <joshfloyd@linkinternational.com>
Subject: Recommendation for Vera Holmes

---

Dear Ms. Curry,

Having been Vera Holmes' supervisor for the last six years, I feel I am in the perfect position to recommend her for the role of sales manager at your company.

> work one's way up 指「逐步晉升」。

Ms. Holmes began as a sales associate at Link International and worked her way up to become a sales development representative. Her duties in this position included **generating**[1] **leads**,[2] **sourcing**[3] and **chasing**[4] client **referrals**,[5] and conducting product training.

> 指「就各方面來說」。

By all accounts, Ms. Holmes is an ambitious, **assertive**,[6] and adaptable salesperson who has contributed greatly to the sales department here. She has consistently **delivered**[7] or **exceeded**[8] her sales targets and maintained excellent **rapport**[9] with her clients. I therefore have no **reservations**[10] recommending Vera.

Should you require any further information, please don't hesitate to contact me.

Regards,

Josh Floyd
Sales Director
Link International

## 中譯

柯瑞小姐您好，

過去六年做為薇拉‧霍姆斯的主管，我認為我非常適合推薦她擔任貴司的業務經理。

霍姆斯小姐在聯通國際公司最初是業務助理，並逐步晉升成為業務開發代表。
她在此職位的職責包含開發客戶、獲得與爭取客戶轉介，以及進行產品訓練。

就各方面而言，霍姆斯小姐是個有抱負、果敢且適應力強的業務人員，
她為我們的業務部貢獻良多。她不斷地實現或超越自己的業績目標，
並與客戶維持絕佳的關係。因此我毫無保留地推薦薇拉。

若您需要任何進一步的資訊，請不吝與我聯繫。

謹致，

賈許‧佛洛伊德
業務協理
聯通國際公司

## 字彙

1. **generate** [ˋdʒɛnəˌret] v. 開發

2. **lead** [lid] n. 客戶

3. **source** [sɔrs] v. （從特定來源）獲得

4. **chase** [tʃes] v. 爭取

5. **referral** [rɪˋfɝəl] n. 轉介

6. **assertive** [əˋsɝtɪv] adj. 果敢的

7. **deliver** [dɪˋlɪvə] v. 實現

8. **exceed** [ɪkˋsid] v. 超越

9. **rapport** [ræˋpɔr] n. 和睦的關係

10. **reservation** [ˌrɛzəˋveʃən] n. 保留

## 延伸學習　推薦信　寫作要點

- **Show how you know the person.**
  表示你如何認識對方。

- **Mention some key points about work/responsibilities.**
  提及和工作或職責有關的要點。

- **Give additional comments on personality and attitude.**
  提供和個性及態度有關的評論。

- **Close with a recommendation.**
  以推薦作結。

# 前同事推薦信

To: Maurice May <mmay@cranstoncorp.com>
From: Sonia Cortez <scortez@D&B.com>
Subject: Recommendation for Clyde Turner

Dear Mr. May,

I have had the pleasure of working alongside Clyde Turner for more than three years at Douglas & Briggs. He and I have **collaborated**[1] on a number of projects and were on the same team for some time. I can happily **vouch**[2] for his **abundant**[3] creativity, positive attitude, **interpersonal**[4] skills, and leadership qualities.

> 為 search engine optimization「搜尋引擎優化」的首字縮略語。

As an invaluable member of the marketing team, Clyde encouraged and helped his colleagues to **shine**[5] while managing to create **impactful**[6] ad campaigns and **copy**[7] of his own. With his expertise in SEO, design background, and innovative thinking, Clyde has been able to increase conversion rates and profits for his clients.

> conversion rate「轉換率」指受網路廣告影響進而購買、註冊的瀏覽者占總廣告點擊人數的比例。

Clyde is reliable, **razor-sharp**,[8] and results-driven. I am confident that he will be an asset to any team he joins.

Sincerely,

Sonia Cortez
Marketing Specialist
Douglas & Briggs

## 中譯

梅伊先生您好，

我很榮幸能與克萊德・透納在道格拉斯與布里格斯公司共事超過三年。他和我合作過若干專案，且有段時間隸屬同一個團隊。我很樂意為他的豐沛創意、積極態度、人際關係技巧與領導特質做保證。

身為行銷團隊的寶貴成員，克萊德鼓勵並協助同事們盡情發揮，自己則同時成功創造出強而有力的廣告活動和宣傳文案。由於克萊德的搜尋引擎優化專長、設計背景與創新思維，他得以提高轉換率，並為客戶增加利潤。

克萊德是個可靠、敏銳且結果導向的人。我有信心他在任何他所加入的團隊中，都將是個寶貴人才。

謹啟，

索妮亞・柯提茲
行銷專員
道格拉斯與布里格斯公司

## 字彙

1. **collaborate** [kəˋlæbəˌret]
   v. 合作；協作

2. **vouch** [vautʃ] v. 證明（+ for）

3. **abundant** [əˋbʌndənt] adj. 豐富的

4. **interpersonal** [ˌɪntəˋpɜsənəl]
   adj. 人際的

5. **shine** [ʃaɪn] v. 表現突出

6. **impactful** [ɪmˋpæktfəl] adj. 有效的

7. **copy** [ˋkɑpi] n. 宣傳說明

8. **razor-sharp** [ˋrezəˋʃɑrp]
   adj. 敏銳的

---

**參考答案**

1. I have had the pleasure of working alongside Sylvia Sims for three years at Rolston Royal.

2. Mark Wheeler began as a marketing assistant at Books for All and worked his way up to become a marketing manager.

3. With her expertise in purchasing, Belinda Carson has been able to improve efficiency at product procurement.

4. I am confident that Brock Hastings will be an asset to the sales team at EVO Corporation.

# Unit 4

# Job Interview Invitations
## 面試通知

## 課前寫作練習

面試通知怎麼寫？
請參考主題詞彙、中譯及括號內的英文提示，將下列句子翻譯成英文。

❶ 非常感謝您應徵克瑞多創意公司（Credo Creations）的業務助理一職。

 _____

_____

主題詞彙

- **attend/attendance**
  [ə`tɛnd] [ə`tɛndəns] 前來
- **confirm** [kən`fɝm]
  確認

- **interest** [`ɪntə͵rɛst]
  興趣
- **interviewee** [͵ɪntəvju`i]
  面試者

❷ 我們已收到您應徵瑪洛科技（Marlowe Technologies）研發工程師一職的資料。

（寫）

_____

_____

❸ 我們很高興邀請您於 3 月 19 日下午兩點到我們公司參加面試。

（寫）

_____

_____

❹ 如果您能在 4 月 10 日前確認是否前來，我們將不勝感激。

（寫）

_____

_____

參考答案請見 p. 55

- **interviewer** [ˋɪntəˏvjuə]
  面試官
- **Q&A session** [ˋsɛʃən]
  問答時間
- **qualification** [ˏkwɑləfəˋkeʃən]
  資格
- **recruitment** [rɪˋkrutmənt]
  招募

TRACK 012

**1** **Thank you very much for your application for the position of (title) with (company).**

非常感謝您應徵（公司）的（職稱）一職。

- **Thank you very much for your application for the position of branch manager with Ralston's Groceries.**
  非常感謝您應徵洛斯頓食品的分店經理一職。

- **Thank you for your interest in the role of (title) with (company).**
  感謝您對（公司）（職稱）一職的興趣。
- **We have received your application for (position) at (company).**
  我們已收到您應徵（公司）（職稱）一職的資料。

---

**2** **It is with great pleasure that we invite you to attend an interview at (location) at (time) on (date).**

我們很高興邀請您於（日期）（時間）到（地點）參加面試。

- **It is with great pleasure that we invite you to attend an interview at our company at 9 a.m. on July 12.**
  我們很高興邀請您於 7 月 12 日上午九點到我們公司參加面試。

- **Your interview has been scheduled for (date), (time) at our company.**
  您的面試已排定於（日期）（時間）在我們公司進行。

# 3 The interview will consist of . . .

面試將包含……

- **The interview will consist of a 40-minute Q&A session with the managers of our various departments.**
  面試將包含與各部門經理的四十分鐘問答時間。

- The interview is scheduled to last approximately (time) and will take the form of . . .
  面試預計歷時約（時間），並將以……的方式進行。
- You will meet with (title), (name), and the interview will last about (time).
  您將與（職稱）（人名）會面，面試將進行約（時間）。

# 4 We would appreciate it if you could confirm your attendance by (date).

如果您能在（日期）前確認是否前來，我們將不勝感激。

- **We would appreciate it if you could confirm your attendance by June 20.**
  如果您能在 6 月 20 日前確認是否前來，我們將不勝感激。

- Please let us know by (date) whether this appointment will be convenient for you.
  請在（日期）以前讓我們知道約定的時間您是否方便。
- Should either the date or time be inconvenient, please contact me at . . .
  如果日期或時間對您來說不方便，請聯絡我……

# 通知面試

| To: | rubyfriedman@delite.com |
|---|---|
| From: | carlsimms@flashfash.org |
| Subject: | Job Interview for Fashion Marketing Director Position |

Dear Ruby:

Thank you very much for your application for the position of fashion marketing director with Flash Fashion. It is with great pleasure that we invite you to attend an interview at our **head office**[1] at 9:30 a.m. on August 25.

The interview will **consist of**[2] a 40-minute **panel**[3] Q&A session with the heads of our **various**[4] departments, followed by a further 20-minute interview with our managing director Joni Huxley. Upon **arrival**,[5] please contact Abigail McCutcheon at **reception**.[6]

We would appreciate it if you could confirm your attendance by August 20.

Many thanks, and we look forward to hearing from you soon.

Sincerely,

Carl Simms
Head of Recruitment
Flash Fashion

## 中譯

露比您好：

非常感謝您應徵閃爍時尚公司的時尚行銷總監一職。我們很高興邀請您於 8 月 25 日上午九點半到我們的總公司參加面試。

面試將包含與各部門主管的四十分鐘小組問答時間，接著再進行與常務董事瓊妮・赫斯利的二十分鐘面試。抵達時，請與接待處的艾比蓋兒・瑪卡奇恩聯繫。

如果您能在 8 月 20 日前確認是否前來，我們將不勝感激。

非常感謝，我們期待很快能收到您的回覆。

謹啟，

卡爾・希姆斯
招聘負責人
閃爍時尚公司

## 字彙

1. **head office** *n.* 總公司

2. **consist of** [kən`sɪst] *v.* 包含

3. **panel** [`pænl] *n.* （選定的）專門小組

4. **various** [`vɛrɪəs] *adj.* 不同的

5. **arrival** [ə`raɪvəl] *n.* 到達

6. **reception** [rɪ`sɛpʃən] *n.* 接待處

### 延伸學習　面試通知 寫作要點

- **Acknowledge receipt of the application.**
  告知已收到職務申請。

- **Give a date, time, and location for the interview.**
  提供面試的日期、時間與地點。

- **State the name of the person the applicant should ask for.**
  說明求職者應該接洽的人。

- **Request confirmation from the applicant.**
  要求求職者回覆確認。

# 婉拒面試

To: waynewilson@spill-net.com
From: hharden@mplt.loc.gov
Subject: Interview Denial for Position of Library Technician

Dear Mr. Wilson:

Thank you for your interest in the position of library technician with Middlestone Public Library Trust. We **regret**[1] to inform you that your application was not successful on this **occasion.**[2] This was **due** in part **to**[3] the **exceptionally**[4] high number of **quality**[5] candidates that applied for the position.

> 指「部分地；在一定程度上」，意同 in some degree。

We very much appreciate your interest in this role with Middlestone Public Library Trust and wish you every personal and professional success with your continued job search.

Many thanks for your time.

Sincerely,

Harry Harden
Human Resources
Middlestone Public Library Trust

威爾森先生您好：

感謝您對中石公共文庫信託公司圖書館管理員一職的興趣。我們很遺憾要通知您，您這次的申請並不成功。部分原因是由於有非常多不錯的候選人申請這份職務。

我們非常感謝您對中石公共文庫信託公司這份職務的興趣，也祝福您在接下來的求職路上獲得個人與專業上的成功。

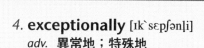

非常感謝您撥冗閱讀。

謹啟，

哈利‧哈登
人力資源部
中石公共文庫信託公司

## 字彙

1. **regret** [rɪˋgrɛt] *v.* 遺憾；惋惜

2. **occasion** [əˋkeʒən] *n.* 時候；時刻

3. **due to** [dju] *prep.* 因為、由於

4. **exceptionally** [ɪkˋsɛpʃənḷi] *adv.* 異常地；特殊地

5. **quality** [ˋkwɑlətɪ] *adj.* 優良的

---

參考答案

1. Thank you very much for your application for the position of sales assistant with Credo Creations.

2. We have received your application for the R & D engineer position at Marlowe Technologies.

3. It is with great pleasure that we invite you to attend an interview at our company at 2 p.m. on March 19.

4. We would appreciate it if you could confirm your attendance by April 10.

## Unit 5

# Interview Thank You Notes
## 面試感謝信

### 課前寫作練習

面試感謝信怎麼寫？
請參考主題詞彙、中譯及括號內的英文提示，將下列句子翻譯成英文。

❶ 感謝您自繁忙的行程中抽空與我進行客服代表一職的面試。

 _____

_____

主題詞彙

- **career** [kəˋrɪr]
  職涯
- **client base**
  客戶群
- **consideration** [kənˌsɪdəˋreʃən]
  考慮
- **expertise** [ˌɛkspɚˋtiz]
  專長；絕技

❷ 我非常興奮有這個機會能與班布里吉通訊公司
（Bainbridge Communications）一同探索可能的
職涯。

⟨寫⟩

_____

_____

❸ 我有自信我身為軟體工程師的專長會是卡頓電子公
司（Cotton Electronics）的資產。

⟨寫⟩

_____

_____

❹ 一如我在面試中所提到的，我相信我的人際關係技
巧符合貴司行銷團隊欲找尋的特質。

⟨寫⟩

_____

_____

參考答案請見 p. 63

- **industry leader**
  業界領袖
- **informative** [ɪnˋfɔrmətɪv]
  增廣見聞的

- **intern** [ɪnˋtɜn]
  實習
- **veteran** [ˋvɛtərən]
  經驗豐富的人；老手

**1** **Thank you for taking time out of your busy schedule to interview me for the (title) position.**

感謝您自繁忙的行程中抽空與我進行（職稱）一職的面試。

- Thank you for taking time out of your busy schedule to interview me for the **product manager** position.

  感謝您自繁忙的行程中抽空與我進行產品經理一職的面試。

 • It was a pleasure meeting you on **June 23** to discuss the **financial analyst** position in your company.

  很高興在 6 月 23 日與您會面，討論貴司財務分析師職缺一事。

---

**2** **I'm very excited about the opportunity to explore a potential career with (company name).**

我非常興奮有這個機會能與（公司名）一同探索可能的職涯。

- I'm very excited about the opportunity to explore a potential career with **Nelson Machinery.**

  我非常興奮有這個機會能與尼爾森機械一同探索可能的職涯。

 • I found it highly edifying to observe how an industry leader like **Samson Tech** approaches **software solution development.**

  看到三孫科技這樣的業界領袖因應軟體解決方案開發的方式，我深獲啟發。

## 3 In our previous discussion, I had the chance to mention my expertise in . . .

在我們先前的討論中，我有機會提到我在……的專長。

- In our previous discussion, I had the chance to mention my expertise in **the development of advertising algorithms.**

  在我們先前的討論中，我有機會提到我在廣告演算法開發上的專長。

 • I am confident that my expertise as **a data administrator** would be an asset to **Jenson Corporation.**

  我有自信我身為數據管理員的專長會是詹森企業的資產。

---

## 4 As I mentioned in the interview, I believe my . . . are in line with . . .

一如我在面試中所提到的，我相信我的……符合……

- As I mentioned in the interview, I believe my **skills as an analyst** are in line with **what your team is searching for.**

  一如我在面試中所提到的，我相信我身為分析師的能力，符合貴司團隊欲找尋的特質。

 • I'd like to add that apart from **being fluent in French,** I am also **CILS** 3 certified in Italian.

  我想補充說明，除了精通法語，我也有義大利文三級檢定的證照。

> 為 Certificazione di Italiano come Lingua Straniera 的縮寫，是義大利語的國家檢定考試。

# 職場新鮮人篇

To: LeBlanc@molenakdistributors.com
From: mmalburke@elite.com
Subject: Appreciation for Interview Offer
Attached: References.docx

Dear Mr. LeBlanc:

Thank you for taking time out of your busy schedule to interview me for the sales assistant position. I'm very excited about the opportunity to explore a **potential**[1] career with Molenak Distributors.

As a recent graduate, I found it informative to **observe**[2] how an industry leader like Molenak **approaches**[3] the **varied**[4] needs of a **diverse**[5] client base in such an innovative way.

> 指「符合；與⋯⋯ 處於同等水準」。

As I mentioned in the interview, I believe my knowledge of business **theory**[6] and interpersonal skills are in line with what your sales team looks for in its members. I've attached references to this e-mail from both my **thesis**[7] advisor and my **superior**[8] at the firm where I interned.

I am very interested in working for you and look forward to hearing from you once your final decision has been reached.

Best,

Matilda Malburke

## 中譯

勒布朗克先生您好：

感謝您自繁忙的行程中抽空與我進行業務助理一職的面試。我非常興奮有這個機會能與莫勒藍克經銷公司一同探索可能的職涯。

剛畢業的我看到莫勒藍克這樣的業界領袖，能以如此創新的方式因應不同客戶群的多樣需求，深覺增廣見聞。

一如我在面試中所提到的，我相信我的商業理論和人際關係技巧知識，符合貴司業務團隊欲找尋成員的特質。我已在信中附上論文指導教授和實習公司主管這兩人的推薦信。

我非常想為貴司效力，期盼在您做出最後結論後，能收到您的通知。

祝一切安好，

瑪緹妲‧麥柏克

## 字彙

1. **potential** [pəˋtɛnʃəl] *adj.* 潛在的

2. **observe** [əbˋzɝv] *v.* 注意到

3. **approach** [əˋprotʃ] *v.* 因應；處理

4. **varied** [ˋvɛrɪd] *adj.* 多樣的

5. **diverse** [daɪˋvɝs] *adj.* 形形色色的

6. **theory** [ˋθiərɪ] *n.* 理論

7. **thesis** [ˋθisəs] *n.* 論文

8. **superior** [suˋpɪrɪə] *n.* 上司

延伸
學習

## 面試感謝信 寫作要點

- Express your appreciation and reinforce your interest in the position.
  表達謝意並強化你對該職務的興趣。

- Suggest how your experience and skills can help with their challenges.
  表明你的經驗和技能能如何為他們的挑戰帶來幫助。

- Say that you're willing to provide any additional information and confirm when a final decision is to be made.
  表示你願意提供任何的額外資訊，並確認何時會有最終決定。

# 職場老鳥篇

To: jung@grafftech.com
From: daviswarburn@mail4days.com
Subject: Appreciation for Interview Offer

Dear Ms. Jung:

It was a pleasure meeting you on June 14 to discuss my possible **addition**[1] to the application development team at Graff Tech.

I've long **aspired**[2] to work at your company as Graff Tech is about developing the best software on the market. In our previous discussion, I had the chance to mention my expertise in the software development field. As a veteran with 15 years of experience in the industry, I am confident that my design and **coding**[3] skills would be an **asset**[4] to Graff Tech.

I'd like to add that apart from working for several major developers, I've also **independently**[5] **released**[6] five successful apps on the online market.

Please feel free to contact me at daviswarburn@mail4days.com if any further information is required or if I am to prepare anything should I be granted a second interview.

I look forward to hearing from you.

Davis Warburn

## 中譯

楊恩女士您好：

很高興在 6 月 14 日與您會面，討論我可能加入葛萊弗科技的應用程式開發團隊一事。

我一直很嚮往能在貴司工作，因為葛萊弗科技所做的就是開發市場上最棒的軟體。在我們先前的討論中，我有機會提到我在軟體開發界的專長。身為擁有十五年業界資歷的老手，我有自信我的設計和編碼能力會是葛萊弗科技的資產。

我想補充說明，除了曾為數家主要開發商效力，我也在線上市場獨力推出過五款成功的應用程式。

若需要進一步的資訊，或因獲得二次面試而有我得準備的任何事項，請隨時寫信至 **daviswarburn@mail4days.com** 與我聯繫。

期盼聽到您的消息。

戴維斯 · 瓦柏恩

## 字彙

1. **addition** [əˋdɪʃən] *n.* 增加的人事物

2. **aspire** [əˋspaɪr] *v.* 有志於；渴望

3. **code** [kod] *v.* 為……編碼

4. **asset** [ˋæˌsɛt] *n.* 資產

5. **independently** [ˌɪndəˋpɛndəntli] *adv.* 獨立地

6. **release** [rɪˋlis] *v.* 推出；發表

---

**參考答案**

1. Thank you for taking time out of your busy schedule to interview me for the customer service representative position.

2. I'm very excited about the opportunity to explore a potential career with Bainbridge Communications.

3. I am confident that my expertise as a software engineer would be an asset to Cotton Electronics.

4. As I mentioned in the interview, I believe my interpersonal skills are in line with what your marketing team is looking for.

# Job Offer Letters
## 錄取通知

## 課前寫作練習

錄取通知怎麼寫？
請參考主題詞彙、中譯及括號內的英文提示，將下列句子翻譯成英文。

**❶** 我們很高興能請您擔任價值領導公司
（**ValuLeader**）的行銷經理一職。

---

**主題詞彙**

- **check in** [tʃɛk]
  報到
- **diploma** [dəˋplomə]
  畢業證書

- **headshot** [ˋhɛdˏʃɑt]
  大頭照
- **hire date** [haɪr]
  聘用日期

❷ 若您接受這份工作，正式的聘用日期將是 11 月 15 日，當天您應於上午九點向安娜・塔勒夫斯基（Anna Taleveski）報到。

寫

❸ 貴賓科技公司（VIP Technologies）全體同仁期待您加入我們的行列。

寫

❹ 身為全職雇員的您將享有醫療與牙科的全額保險。

寫

參考答案請見 p. 71

- **medical examination**
  [ɪgˌzæməˋneʃən] 體檢

- **passbook** [ˋpæsˌbʊk]
  銀行存摺

- **starting salary** [ˋsæləri]
  起薪

- **termination certificate**
  [səˋtɪfɪkət] 終止聘雇證明

## 1 We are pleased to offer you the position as our (title) at (company name).

我們很高興能請您擔任（公司名）的（職稱）一職。

- **We are pleased to offer you the position as our budget director at Global Accounting.**
  我們很高興能請您擔任全球會計事務所的預算協理一職。

 • Please accept this letter as confirmation of our offer made to you for the position of **senior engineer** in our **MIS** department.
請同意此信為確認我們提供您敝司資訊管理部的資深工程師一職。

---

## 2 If you accept this offer, your official hire date will be (date), on which day you should check in with (name) at (time).

若您接受這份工作，正式的聘用日期將是（日期），當天您應於（時間）向（人名）報到。

- **If you accept this offer, your official hire date will be October 15, on which day you should check in with Leah Feddler at 8:30 a.m.**
  若您接受這份工作，正式的聘用日期將是 10 月 15 日，當天您應於上午八點半向莉雅・費德勒報到。

 • The starting date is **January 10;** your starting salary will be **$3,000 per month, which will be paid on the 10th of every month.**
到職日為 1 月 10 日，您的起薪將是每月三千元，發薪日則是每月 10 日。

## 3 We at (company name) look forward to welcoming you aboard.

（公司名）全體同仁期待您加入我們的行列。

- **We at Lively Textile look forward to welcoming you aboard.**

  萊服利紡織全體同仁期待您加入我們的行列。

 • As a full-time employee, you will be **eligible** for **labor and health insurance.**

身為全職雇員的您將享有勞健保。

> 讀做 [ˈɛlədʒəbəl]，指「有資格的」。

- You are entitled to **a week of paid holidays** during your first year of employment.

  第一年雇用期間，您會有一星期的給薪年假。

---

## 4 Feel free to contact (name) by e-mail if you have any additional questions or concerns.

若您有任何其他的問題或疑慮，歡迎透過電子郵件與（人名）聯絡。

- **Feel free to contact our HR manager, Bill Blakely, by e-mail if you have any additional questions or concerns.**

  若您有任何其他的問題或疑慮，歡迎透過電子郵件與我們的人資經理比爾‧布雷克利聯絡。

 • Please let us know at your earliest convenience whether **you will be taking up this offer.**

請盡早讓我們知道您是否會接下這份工作。

# 聘雇信函

---

To: hrothbart@jobs4U.com
From: arandleman@freemantechnologies.com
Subject: Job Offer for Sales Representative

---

Dear Ms. Rothbart:

We are pleased to offer you the position as our sales representative at Freeman Technologies. Your starting salary will be the **figure**[1] we previously agreed upon. If you accept this offer, your official hire date will be September 4, on which day you should check in with Gideon Bairnez at 9 a.m.

On this date, please bring the following with you:
- The original termination certificate from your **previous**[2] employer
- A copy of your diploma
- A copy of your ID card
- A copy of your Taipei Trust banking passbook
- A 2 x 2-inch headshot photo
- The original **documents**[3] of a recent medical examination report

We at Freeman Technologies look forward to welcoming you **aboard**.[4] Feel free to contact me by e-mail if you have any additional questions or concerns.

Best,

Ari Randleman
HR Director
Freeman Technologies

## 中譯

羅斯巴特女士您好：

我們很高興能請您擔任費里曼科技的業務代表一職。您的起薪將如同先前我們所同意的數字。若您接受這份工作，正式的聘用日期將是 **9** 月 **4** 日，當天您應於上午九點向吉迪恩‧貝爾涅茲報到。

當天請攜帶下列文件：

• 先前任職公司的終止聘雇證明正本
• 畢業證書影本一份
• 身分證影本一份

• 台北信託銀行存摺影本一份
• 兩吋大頭照一張
• 近期體檢報告的正本文件

費里曼科技全體同仁期待您加入我們的行列。若您有任何其他的問題或疑慮，歡迎透過電子郵件與我聯絡。

祝一切安好，

阿里‧蘭德曼
人資協理
費里曼科技

## 字彙

1. **figure** [ˋfɪgjə] n. 數字

2. **previous** [ˋpriviəs] adj. 先前的

3. **document** [ˋdɑkjəmənt] n. 文件

4. **aboard** [əˋbɔrd] adv. 加入行列地

---

延伸
學習

## 錄取通知 寫作要點

- Clearly state that you are making an offer of employment.
  清楚表明你要聘雇對方。

- List details about the job, including the starting date and starting salary.
  列出工作相關細節，包括到職日和起薪。

- Request that the recipient acknowledge the offer in writing.
  要求收信者回信告知是否接受該職務。

# 回覆信函

To: arandleman@freemantechnologies.com
From: hrothbart@jobs4U.com
Subject: Re: Job Offer for Sales Representative

Dear Mr. Randleman:

I am very pleased to accept the sales representative position with Freeman Technologies. Thank you for making the interview process **straightforward**[1] and **manageable**.[2] I am eager to make a positive contribution to the company and work **alongside**[3] everyone on the team.

I feel that we will have a great working relationship in the future and that I will weave into the culture at Freeman Technologies very well.

> weave into [wiv] 指「融入……」。

I look forward to starting employment on September 4 at 9 a.m. Should there be any additional information or paperwork you need **prior to**[4] then, please do not hesitate to contact me.

Again, thank you very much.

Regards,

Hannah Rothbart

## 中譯

蘭德曼先生您好：

我很高興接受費里曼科技的業務代表一職。感謝您讓面試的過程簡單且順利。
我非常想為貴司做出積極的貢獻，並與團隊的每一位同仁並肩合作。

我覺得未來我們會有很棒的工作關係，而我也將完全融入費里曼科技的文化。

我期待在 9 月 4 日上午九點開始工作。在那之前，
若您需要任何其他的資訊或文件，請不吝與我聯絡。

再次非常感謝您。

謹致，

漢娜‧羅斯巴特

## 字彙

1. **straightforward** [ˌstretˋfɔrwəd]
   adj. 簡單的

2. **manageable** [ˋmænɪdʒəbəl]
   adj. 可應付的

3. **alongside** [əˋlɔŋˋsaɪd]
   adv. 與⋯⋯一起

4. **prior to** [praɪr] prep. 在⋯⋯之前

參考答案

1. We are pleased to offer you the position as our marketing manager at ValuLeader.

2. If you accept this offer, your official hire date will be November 15, on which day you should check in with Anna Taleveski at 9 a.m.

3. We at VIP Technologies look forward to welcoming you aboard.

4. As a full-time employee, you will be eligible for full medical and dental coverage.

# Unit 7 Job Rejection Letters
## 未錄取通知

## 課前寫作練習

未錄取通知怎麼寫？
請參考主題詞彙、中譯及括號內的英文提示，將下列句子翻譯成英文。

❶ 感謝您抽空與我們會面，洽談南方科技公司
（SouthTech Corp.）的資深工程師一職。

（寫）

主題詞彙

- **candidacy** [ˋkændədəsi]
  候選人資格
- **competitive**
  [kəmˋpɛtətɪv] 競爭激烈的

- **credential** [krɪˋdɛnʃəl]
  資歷
- **opening** [ˋopənɪŋ]
  職缺

**❷** 很遺憾地通知您，芙莉蒙特貿易公司（Freemont Trading）將不會考量您為此職缺的人選。

（寫）

---

---

**❸** 雖然您擁有行銷方面的傑出成就，但我們的甄選過程競爭相當激烈，而我們已改選擇另一位候選人了。

（寫）

---

---

**❹** 再次感謝您表達對全球禮品公司（WorldWide Gifts）的興趣，並祝您未來求職一切順利。

（寫）

---

---

參考答案請見 p. 79

---

- **opportunity** [ˌɑpəˈtjunəti]
  機會；可能性
- **profile** [ˈproˌfaɪl]
  傳略；人物簡介

- **qualification** [ˌkwɑləfəˈkeʃən]
  資格
- **selection process**
  [səˈlɛkʃən] 甄選過程

## 1 We would like to thank you for taking the time to meet with us about the (title) role at (company name).

感謝您抽空與我們會面，洽談（公司名）的（職稱）一職。

- We would like to thank you for taking the time to meet with us about the **security expert** role at **Gainly Freeling.**

  感謝您抽空與我們會面，洽談甘利福林公司的保全專家一職。

 • We appreciate your interest in the **assistant** position at **LiveVille.**

  感謝您對生活村公司助理一職的興趣。

## 2 We regret to inform you that (company name) will not be pursuing your candidacy for this position.

很遺憾地通知您，（公司名）將不會考量您為此職缺的人選。

- We regret to inform you that **Padrox Inc.** will not be pursuing your candidacy for this position.

  很遺憾地通知您，帕德洛克斯公司將不會考量您為此職缺的人選。

 • Unfortunately, after careful consideration, we have determined that we will not be pursuing your application.

  遺憾的是，幾經深思熟慮，我們決定不進行您的聘雇作業。

## 3 Although you possess . . ., our selection process was highly competitive, and we have chosen to move forward with another candidate.

雖然您擁有⋯⋯，但我們的甄選過程競爭相當激烈，而我們已改選擇另一位候選人了。

- Although you possess **extensive work experience, our selection process was highly competitive, and we have chosen to move forward with another candidate.**

  雖然您擁有豐富的工作經驗，但我們的甄選過程競爭相當激烈，而我們已改選擇另一位候選人了。

 相關寫法
- While we were impressed by your **education and work experience,** we ended up selecting another candidate whose abilities we believe **would better suit the position.**

  儘管我們對您的教育背景與工作經驗印象深刻，但我們最後選擇了另一位候選人，我們相信他的能力較符合此職缺。

## 4 We thank you again for expressing interest in (company name) and wish you all the best in your future endeavors.

再次感謝您表達對（公司名）的興趣，並祝您未來求職一切順利。

- We thank you again for expressing interest in **Bilton Automotives** and wish you all the best in your future endeavors.

  再次感謝您表達對畢爾頓汽車的興趣，並祝您未來求職一切順利。

 相關寫法
- There is no doubt that you will have success in **finding suitable employment.**

  毫無疑問地，您一定能順利找到適合的工作。

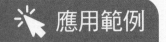

# 拒聘信函

To: anewalla@jobs.com
From: sroti@ghsic.com
Subject: Denial of Job Position

Dear Mr. Newalla:

We would like to thank you for taking the time to meet with us about the financial consultant role at Guiding Hand Securities Investment Consulting. We regret to **inform**[1] you that Guiding Hand will not be **pursuing**[2] your candidacy for this position.

Although you **possess**[3] impressive qualifications, our selection process was highly competitive, and we have chosen to move forward with another candidate whose credentials better meet the company's needs at this time. However, as you were a strong applicant, we will be keeping your résumé on file for future openings that are more suitable for your profile and abilities.

We thank you again for expressing interest in Guiding Hand and wish you all the best in your future **endeavors**.[4]

Best,

Shaun Roti
HR Director
Guiding Hand Securities Investment Consulting

## 中譯

努瓦拉先生您好：

感謝您抽空與我們會面，洽談指導之手證券投資顧問公司的財務顧問一職。很遺憾地通知您，指導之手將不會考量您為此職缺的人選。

雖然您擁有令人印象深刻的條件，但我們的甄選過程競爭相當激烈，而我們已改選擇另一位候選人，他的資歷較符合公司現階段的需求。然而，由於您是一位優秀的申請者，我們會將您的履歷歸檔，用於未來更適合您資歷與能力的職缺。

再次感謝您表達對指導之手的興趣，並祝您未來求職一切順利。

祝一切安好，

尚恩・洛提
人資協理
指導之手證券投資顧問公司

## 字彙

1. **inform** [ɪnˋfɔrm] *v.* 通知

2. **pursue** [pɚˋsu] *v.* 繼續進行

3. **possess** [pəˋzɛs] *v.* 擁有；具有

4. **endeavor** [ɪnˋdɛvɚ] *n.* 努力

---

### 延伸學習  拒聘函 寫作要點

- **Thank the candidate for their interest and for attending the interview.**
  感謝求職者展現出興趣並參加面試。

- **Tactfully explain why you were rejecting the candidate.**
  婉轉解釋拒絕該求職者的原因。

- **To make the letter less personal, use "we" instead of "I" throughout.**
  全篇使用「我們」而非「我」的稱謂來使信件語氣較公正。

# 回覆信函

To: sroti@ghsic.com
From: anewalla@jobs.com
Subject: Re: Denial of Job Position

Dear Mr. Roti:

Thank you so much for giving me the opportunity to interview for the financial consultant opening at Guiding Hand Securities Investment Consulting. I appreciate you taking time out of your busy schedule to introduce your company and discuss the position with me.

> 指「為了該目的」。

I would very much appreciate being considered for future positions with your company. To that end, I would like to **reach out**[1] to you and ask for any suggestions or **constructive**[2] **criticism**[3] you might have in regard to my résumé or interview performance. I **value**[4] your opinion, and I would like to make sure that I put my best foot forward in the future.

Regards,

> put one's best foot forward 指「全力以赴」。

Auston Newalla

中譯

洛提先生您好：

非常感謝您給我這個機會，面試指導之手證券投資顧問公司的財務顧問一職。
謝謝您在百忙之中抽空介紹貴司，並與我談論此職缺。

針對貴司未來的職缺，若我能被納入考慮，
我將感激不盡。為此目的，我想請求您的協
助，徵詢您對於我的履歷或面試表現而可能
有的任何建議或建設性的批評。我相當重視
您的意見，而且我想確定自己未來能表現出
最好的一面。

謹致，

奧斯頓・努瓦拉

## 字彙

1. **reach out** *v.* 向……伸出援手

2. **constructive** [kən`strʌktɪv]
   *adj.* 建設性的

3. **criticism** [`krɪtə͵sɪzəm] *n.* 批評

4. **value** [`vælju] *v.* 重視

參考答案

1. We would like to thank you for taking the time to meet with us about the senior engineer role at SouthTech Corp.

2. We regret to inform you that Freemont Trading will not be pursuing your candidacy for this position.

3. Although you possess excellent achievements in marketing, our selection process was highly competitive, and we have chosen to move forward with another candidate.

4. We thank you again for expressing interest in WorldWide Gifts and wish you all the best in your future endeavors.

# Unit 8

# Notice of Personnel Changes
## 人事異動通知

## 課前寫作練習

人事異動通知怎麼寫？
請參考主題詞彙、中譯及括號內的英文提示，將下列句子翻譯成英文。

❶ 這封電郵旨在通知大家，公司財務和會計部的重要變動。

_____

_____

主題詞彙

- **promote** [prə`mot]
  晉升
- **demote** [dɪ`mot]
  降職

- **incoming** [`ɪn͵kʌmɪŋ]
  新任的
- **outgoing** [`aut͵goɪŋ]
  即將離職的

❷ 自 7 月 1 日起，克萊兒 · 貝蒙特（Claire Belmont）
將晉升為文字編輯。

㊢
_____

_____

❸ 因為葛哈德 · 富勒（Gerhard Fuller）日前辭職，
賽 · 史戴特勒（Sy Stetler）將接替為區域經理。

㊢
_____

_____

❹ 自 8 月 1 日起，葛蕾西 · 麥坎（Gracie Malcom）
將從研發部轉調至資訊科技部。

㊢
_____

_____

參考答案請見 p. 87

- **transfer** [ˈtræns͵fɝ]
  調職
- **retire** [rɪˈtaɪr]
  退休

- **predecessor** [ˈprɪdə͵sɛsə]
  前任
- **successor** [səkˈsɛsə]
  繼任者

## 1 This e-mail is to inform you of important changes in our (type) departments.

這封電郵旨在通知大家，公司（類別）部門的重要變動。

- This e-mail is to inform you of important changes in our **sales and marketing** departments.

  這封電郵旨在通知大家，公司業務和行銷部的重要變動。

- We are writing to inform you of personnel changes within the company.

  來信旨在告知公司內部的人事變動。

- This is to inform all employees of recent staff changes at Joaquin Academic Institution.

  此信旨在通知所有瓦昆學術機構的員工最近的人員異動。

## 2 Effective (date), (name) will be promoted . . .

自（日期）起，（人名）將晉升為……

- Effective **May 31, Erica Jones** will be promoted **to senior financial analyst**.

  自 5 月 31 日起，艾芮卡・瓊斯將晉升為資深財務分析師。

- **Eugene Moore** has been promoted to **production manager** in the **manufacturing** department.

  尤金・摩爾已被拔擢為製造部的產線經理。

## 3 In light of (A)'s recent resignation, (B) will be taking over as (title).

因為 A 日前辭職，B 將接替為（職稱）。

- In light of **Kate Benning**'s recent resignation, **Dolores Hughes** will be taking over as **corporate social responsibility manager**.

  因為凱特・班寧日前辭職，朵洛莉絲・休斯將接替為企業社會責任經理。

- The new role of **director of Ethics and Compliance** will be filled by **Scott Gerard**.

  道德與合規協理的新職位將由史考特・傑拉德遞補。

- **Frank Albrecht** has replaced **Kate Merowe** as **the new district manager**.

  法蘭克・歐柏萊克特已接替凱特・梅洛為新任區域經理。

---

## 4 As of (date), (name) will be transferring from (Department A) to (Department B).

自（日期）起，（人名）將從（部門 A）轉調至（部門 B）。

- As of **March 1**, **Harry Lu** will be transferring from **R & D** to **the quality assurance department**.

  自 3 月 1 日起，哈利・盧將從研發部轉調至品保部。

- **Auditing** will now be in the capable hands of **Maria Dominguez**, who is transferring from **Accounting** to become **chief audit executive**.

  查帳將由瑪莉亞・多明格茲負責，她從會計部轉調過來成稽核長。

# 升職篇

To: all@absinternainternational.com
From: bsantos@absinternainternational.com
Subject: Leadership Changes in Finance and IT

Dear All,

This e-mail is to inform you of important leadership changes in our finance and IT departments. **Effective**[1] January 15, two staff members will be promoted as part of our **revised**[2] corporate **mission**.[3]

After 10 years of loyal service as a senior financial analyst, Terézia Arnoni will now **liaise**[4] with our global finance **division**[5] in her new role as investment manager.

Thirty-five years after he first began his career with Absinterna, Lou Miles will be **stepping down**[6] from his position as global head of IT Compliance. He will be **succeeded**[7] by his **protégé**,[8] Galena Beaumont, who has worked under Lou for the past 17 years.

Please join me in wishing Lou well in his retirement and in congratulating Terézia and Galena on their new roles.

Regards,

Beyza Santos
HR Director
Absinterna International

## 中譯

大家好，

這封電郵旨在通知大家，公司財務部和資訊科技部的重要領導階層變動。自 1 月 15 日起，做為調整公司企業使命的一部分，兩名職員將獲得升遷。

以資深財務分析師的身分忠誠服務了十年後，泰芮西亞‧阿隆尼現在將以投資經理的新角色來協調公司的全球財務事業處。

在艾柏辛特納公司展開職涯的三十五年後，路‧邁爾斯將卸下資訊科技合規事業處的全球主管一職。繼任者將是他的門生蓋琳娜‧波蒙，她過去十七年一直在路的底下做事。

請和我一同祝福路的退休生活如意，並對泰芮西亞和蓋琳娜的新角色予以祝賀。

謹致，

貝莎‧桑托斯
人資協理
艾柏辛特納國際公司

## 字彙

1. **effective** [ɪˋfɛktɪv] *adj.* 生效的

2. **revised** [rɪˋvaɪzd] *adj.* 調整的

3. **mission** [ˋmɪʃən] *n.* 使命

4. **liaise** [liˋez] *v.* 協調；聯繫

5. **division** [dəˋvɪʒən] *n.* 事業處

6. **step down** *v.* 退休；退位

7. **succeed** [səkˋsid] *v.* 接替

8. **protégé** [ˋprotəˌʒe] *n.* 門生

## 延伸學習　升遷公告 補充句型

- (Name) has been promoted to (position) in (department).
  （人名）已被拔擢為（部門）的（職稱）。

- After (number) years in (department), (name) will head up ...
  在（部門）任職（數字）年後，（人名）將帶領……

- (Name) has been designated as (title) of (department), effective (date).
  自（日期）起，（人名）被指派為（部門）的（職稱）。

# 職務調動篇

To: all@absinternainternational.com
From: bsantos@absinternainternational.com
Subject: Personnel Changes

Dear All,

We are writing to inform you of **personnel**[1] changes within the company. The key **organizational**[2] moves we're making to help prepare us for the opportunities and challenges ahead include the following:

> 指「因為；鑒於」，意同 in view of。

- In light of Rachel Schulman's recent **resignation**,[3] Sandi Kravitz will be taking over as product manager.

- As of January 30, Dick Simmons will be transferring from Sales to Marketing.

- Due to growing **regional**[4] sales, Burkhart Smith will continue his work as senior marketing manager as a new member of the Asia/Oceana team in International Marketing.

- Operations will now be in the capable hands of Matthew McCrory, who is transferring from the finance department to become head of Operations.

- The new role of director of Customer Experience will be filled by Rajani Kaur.

We hope the changes announced today will help us further develop our talent and help our business **evolve**.[5]

Best,

Beyza Santos
HR Director
Absinterna International

## 中譯

大家好，

來信旨在告知公司內部的人事變動。為幫助我們準備好迎接未來機會和挑戰所做的重要組織變動如下：

- 因為瑞秋‧舒曼日前辭職，珊蒂‧克羅維茲將接替為產品經理。
- 自 1 月 30 日起，迪克‧西蒙斯將從業務部轉調至行銷部。
- 由於區域業務的成長，柏克哈特‧史密斯將以國際行銷部亞洲／大洋洲團隊新成員的身分，繼續他資深行銷經理的工作。
- 營運將由馬修‧麥可洛里負責，他從財務部轉調過來成營運部主管。
- 客戶體驗部協理的新職位將由勒嘉妮‧寇爾遞補。

我們希望今日所公布的變動能助我們進一步發展公司的人才，並讓事業蒸蒸日上。

祝一切安好，

貝莎‧桑托斯
人資協理
艾柏辛特納國際公司

## 字彙

1. **personnel** [ˌpɜsəˈnɛl] *n.* 人事

2. **organizational** [ˌɔrɡənəˈzeʃən] *adj.* 組織的

3. **resignation** [ˌrɛzɪɡˈneʃən] *n.* 辭職

4. **regional** [ˈridʒənl] *adj.* 區域的

5. **evolve** [ɪˈvɑlv] *v.* 逐步發展

---

**參考答案**

1. This e-mail is to inform you of important changes in our finance and accounting departments.

2. Effective July 1, Claire Belmont will be promoted to copy editor.

3. In light of Gerhard Fuller's recent resignation, Sy Stetler will be taking over as regional manager.

4. As of August 1, Harry Lu will be transferring from R & D to the IT department.

# Resignation Letters

## 辭職信

## 課前寫作練習

辭職信怎麼寫？
請參考主題詞彙、中譯及括號內的英文提示，將下列句子翻譯成英文。

**❶** 此信旨在正式通知您，我將自 3 月 1 日起，辭去數據分析師一職。

---

---

主題
詞彙

- **departure** [dɪˋpɑrtʃə]
  離開（公司）
- **effective** [ɪˋfɛktɪv]
  生效的

- **employment** [ɪmˋplɔɪmənt]
  受雇；就業
- **gratitude** [ˋɡrætəˏtud]
  感謝；感激之情

❷ 謹正式通知，我將於 6 月 30 日卸任史密斯公司
（Smith Company）的行銷主管一職。

(寫) _____

_____

❸ 我提早兩個月遞交辭呈，因為我知道我的接替者將
需要大量的培訓。

(寫) _____

_____

❹ 我在寇曼有限公司（Coldman Ltd.）的工作資歷，
賦予我寶貴的商業洞見。

(寫) _____

_____

參考答案請見 p. 95

- **notify** [ˋnotəˏfaɪ]
  （正式）通知
- **replacement** [rɪˋplesmənt]
  接替者

- **resign** [rɪˋzaɪn]
  辭職
- **transition** [trænˋzɪʃən]
  交接

**1** **The purpose of this letter is to officially notify you of my resignation from my position as (title) effective (date).**

此信旨在正式通知您，我將自（日期）起，辭去（職務）一職。

- The purpose of this letter is to officially notify you of my resignation from my position as **district manager** effective **March 12.**

  此信旨在正式通知您，我將自 3 月 12 日起，辭去區域經理一職。

 • I'm writing to give my formal notice that I'll be leaving my role as **office manager** for **Lamar Advertising** on **July 31.**

謹正式通知，我將於 7 月 31 日卸任拉馬爾廣告公司的行政經理一職。

---

**2** **I am tendering my resignation (time) in advance because . . .**

我提早（時間）遞交辭呈，因為……

- I am tendering my resignation **one month** in advance because **I wish to provide ample time for a suitable replacement to be found.**

  我提早一個月遞交辭呈，因為我希望能提供足夠的時間來找到合適的接替者。

 • This notification provides you with **three weeks'** notice to plan for my replacement.

此辭職通知讓您有三個星期來計畫我的接替人選。

## 3 My employment at (company name) has granted me . . .

我在（公司名）的工作資歷，賦予我⋯⋯

- **My employment at Reynard's Fine Clothing has granted me invaluable business insight and customer service skills.**

  我在雷納高級布料公司的工作資歷，賦予我寶貴的商業洞見與客服技巧。

- I've learned so much about **marketing strategy and the digital media space,** which I will certainly take with me throughout my career.

  我在行銷策略和數位媒體領域所學甚多，我定會在我的職涯中善加利用。

- Thank you for your support during my **six years** with **Capitol Company.**

  感謝您在我任職凱比多公司的六年期間所給予的支持。

## 4 I intend to make the transition of my responsibilities . . .

我想⋯⋯交接職務。

- **I intend to make the transition of my responsibilities as seamless as possible.**

  我想盡可能完美地交接職務。

- During my last **two weeks,** I'll do everything possible to wrap up my duties and train other team members.

  在我最後的兩週內，我將竭盡全力完成我的職責並培訓其他團隊成員。

# 辭職信函

To: fkillian@coalmangrundy.com
From: trichards@coalmangrundy.com
Subject: Job Resignation

Dear Mr. Killian:

The purpose of this letter is to officially notify you of my resignation from my position as an international sales representative effective March 20.

I am **tendering**[1] my resignation one month in advance because I have received a full **scholarship**[2] to complete my PhD in Munich, Germany.

I am grateful to have been given the opportunity to perform such an enjoyable and **edifying**[3] **vocation**[4] for five years. My employment at Coalman & Grundy has granted me invaluable **hands-on**[5] experience, **imparting**[6] interpersonal and commercial skills which I will utilize for the remainder of my professional career.

I intend to make the transition of my responsibilities as smooth as possible and will ensure that all of my reports are completed before my departure. I am also available to assist in the training of my replacement.

I wish you all the best in the future. I can be contacted by e-mail at tomerrichards@email4days.com.

Thank you again for the opportunity,

Tomer Richards
International Sales Representative
Coalman & Grundy

## 中譯

奇里恩先生您好：

此信旨在正式通知您，我將自 **3 月 20 日**起，辭去國外業務一職。

我提早一個月遞交辭呈，因為我已獲得在德國慕尼黑攻讀博士學位的全額獎學金。

我很感激有這個機會能從事如此愉快且具啟發性的工作達五年的時間。我在柯曼與格蘭迪公司的工作資歷，賦予我寶貴的實務經驗，也給予我人際與商務技能，我會在接下來的職涯善加利用。

我想盡可能順利地交接職務，並確保我所有的報告在我離去前都會完成。我也能協助訓練我的接替者。

祝福您的未來一切順利。您可透過 **tomerrichards@email4days.com** 的電子信箱聯絡我。

再次謝謝您給予的機會，

托莫‧理查德斯
國外業務
柯曼與格蘭迪公司

## 字彙

1. **tender** [ˋtɛndə] *v.* 提交

2. **scholarship** [ˋskɑlə͵ʃɪp] *n.* 獎學金

3. **edifying** [ˋɛdə͵faɪɪŋ] *adj.* 具啟發性的

4. **vocation** [voˋkeʃən] *n.* 職業、工作

5. **hands-on** [ˋhændzˋɑn] *adj.* 實務的

6. **impart** [ɪmˋpɑrt] *v.* 給予

## 延伸學習　辭職信 寫作要點

- **Subject 主旨**
  點出職務名稱和離職生效日期。

- **Notification 通知**
  說明提前多早遞交辭呈和辭職的原因。

- **Gratitude 謝詞**
  感謝公司給予的機會和幫助。

- **Transition 交接**
  表明會妥善處理交接事宜。

- **Closing 結語**
  祝福收信者並提供離職後的聯絡資訊。

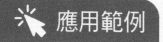 

# 回覆信函

To: trichards@coalmangrundy.com
From: fkillian@coalmangrundy.com
Subject: Re: Job Resignation

Dear Mr. Richards:

This letter is to **confirm**[1] the **approval**[2] of your resignation, **received**[3] on February 20.

We appreciate your early notice as well as your years of hard work and **commitment**[4] to the company. Please **assist**[5] in the training of your replacement, Maria Aldrich, from February 22 to March 20.

Coalman & Grundy would like to thank you for your service and wish you the best of luck in your future endeavors.

Best,

Fred Killian
Sales Director
Coalman & Grundy

> 用來祝願對方未來能成功，常用句型如下：
> • wish sb (the) best of luck in ...
> • best of luck to sb in ...

## 中譯

理查德斯先生你好：

這封信是要確認批准我在 2 月 20 日所收到的辭呈。

我們感謝你的提早通知，也謝謝你多年來的辛勤以及對公司的奉獻。
請於 2 月 22 日至 3 月 20 日期間，協助訓練你的接替者瑪莉亞·歐德里奇。

柯曼與格蘭迪公司感謝你的效勞，並祝你
前途一帆風順。

祝一切安好，

佛萊德·奇里恩
銷售總監
柯曼與格蘭迪公司

## 字彙

1. **confirm** [kən`fɜm] *v.* 確認

2. **approval** [ə`pruvəl] *n.* 批准；同意

3. **receive** [rɪ`siv] *v.* 收到

4. **commitment** [kə`mɪtmənt] *n.* 奉獻

5. **assist** [ə`sɪst] *v.* 協助

---

參考答案

1. The purpose of this letter is to officially notify you of my resignation from my position as data analyst effective March 1.

2. I'm writing to give my formal notice that I'll be leaving my role as marketing supervisor for Smith Company on June 30.

3. I am tendering my resignation two months in advance because I know my replacement will require extensive training.

4. My employment at Coldman Ltd. has granted me invaluable business insight.

# Job Handover Notes
## 離職交接通知

## 課前寫作練習

離職交接通知怎麼寫？
請參考主題詞彙、中譯及括號內的英文提示，將下列句子翻譯成英文。

❶ 記住你應該在每星期一早上提交週報告。

 ........................................................................

........................................................................

主題
詞彙

- **assignment** [əˋsaɪnmənt]
  （分派的）工作
- **consult** [kənˋsʌlt]
  請教；諮詢
- **duty** [ˋdjutɪ]
  職責
- **hand over**
  移交

❷ 有關差旅補貼的問題可轉給會計部的南希
（Nancy）。

寫 _____

_____

❸ 前兩週我一直在訓練我的接替人員約翰，而他已掌
握我們的物流管理作業。

寫 _____

_____

❹ 我已做了必要的安排，將我所有的職責移交給我的
接替者莎拉‧布萊克（Sarah Black）。

寫 _____

_____

參考答案請見 p. 103

- **instruct** [ɪnˋstrʌkt]
  指示
- **replacement** [rɪˋplesmənt]
  接替人員

- **take over**
  接手
- **workload** [ˋwɝkˏlod]
  工作量

## 1 I'd like to jot down some important things to remember for when you . . .

我想趕快寫下你……時得記住的一些重要事項。

- I'd like to jot down some important things to remember for when you **assume the responsibility for hotel booking.**

  我想趕快寫下你承接飯店訂房的職責時得記住的一些重要事項。

 • I want to put down in writing some things you'll need to recall when **you're on your own.**

別忘了這份報告的數據得寄給會計部。
我想寫下來你接手後得記住的一些事情。

---

## 2 Keep in mind that you should . . .

記住你應該……

- Keep in mind that you should **submit all the financial reports to Joyce, our manager, for approval.**

  記住你應該提交所有的財務報告給我們的經理喬伊絲以供批准。

 • Don't forget that **the numbers of this report** should be **e-mailed to Accounting.**

  別忘了這份報告的數據得寄給會計部。

## 3 Questions about (work) can be directed to (name) in (department).

有關（工作）的問題可轉給（部門）的（人名）。

- **Questions about accounts receivable can be directed to Brenda in Finance.**
  有關應收帳款的問題可轉給財務部的布蘭達。

 • If you have any problems concerning **logistics,** you can consult **Simon** in the **shipping** department.
如果你有任何關於物流的問題，可以請教運輸部的賽門。

---

## 4 I've been training my replacement, (name), for (period of time), and he's/she's already mastered (work).

（一段時間）我一直在訓練我的接替人員（人名），而他／她已掌握（工作）。

- I've been training my replacement, **Kate,** for **the last three weeks,** and she's already mastered **our ERP system.**
  前三週我一直在訓練我的接替人員凱特，而她已掌握我們的企業資源計畫系統。

 • My replacement, **Terry,** has been getting trained for **the past few weeks** — he's pretty much up to speed now.
我的接替人員泰瑞過去幾週都在受訓，他現在蠻進入狀況了。

• I have made the necessary arrangements of handing over all my duties to **my replacement, Richard Brown.**
我已做了必要的安排，將我所有的職責移交給我的接替者理查‧布朗。

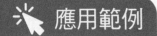應用範例

TRACK 031

# 交接人員通知

To: bbennett@mediaquest.com
From: ssmythe@mediaquest.com
Subject: Key Points about the Job
Attached: Contact Info.xlsx

Dear Bruce:

> 指「永久地」，意同 forever、permanently。

Before I leave for good today, I'd like to **jot down**[1] some important things to remember for when you take over the job.

So far, you've been **picking up**[2] the duties of being an editorial assistant really quickly. Keep in mind that you should give all **proofread**[3] articles to Jill, the content manager, while questions about ads can be directed to Brandon in Marketing. And if your workload ever becomes too much, don't be afraid to let Beth know.

You can use the old issues of the magazine — located in your file cabinet — for **reference**.[4] Also, I've attached the contact info of all our coworkers and **freelancers**[5] here.

If you have any questions after I'm gone, just **shoot**[6] me an e-mail at samsmythe33@hotmail.com.

Good luck with everything, and don't forget that the travel article needs to be finished by noon on Monday!

Best,

Sam

## 中譯

布魯斯你好：

在我今天正式離開以前，我想趕快寫下你接手這份工作時得記住的一些重要事項。

到目前為止，你在熟習編輯助理職責的這部分上手得很快。記住你應該把所有校對過的文章給內容管理員吉兒，而有關廣告的問題可轉給行銷部的布蘭登。倘若你的工作量變得太多，別害怕讓貝絲知道。

你可以使用放在你檔案櫃裡的過期雜誌做為參考。還有，我也在信裡附上所有同事與特約作者的聯絡資訊。

我離職後，如果你還有問題，寄封信到 samsmythe33@hotmail.com 這個電子信箱給我就可以了。

祝你一切順利，也別忘了那篇旅遊文章得在星期一中午前完成！

祝一切安好，

山姆

## 字彙

1. **jot down** [dʒɑt] *v.* 速記

2. **pick up** *v.* 學習；增進

3. **proofread** [ˋprufˏrid] *v.* 校對

4. **reference** [ˋrɛfərəns] *n.* 參考

5. **freelancer** [ˋfriˏlænsə] *n.* 特約作者

6. **shoot** [ʃut] *v.* 快速寄出

---

### 延伸學習　交接信 寫作要點

- Detailed information on your daily activities, tasks, and priorities
  詳細說明你的日常活動、任務及優先事項

- An outline of the key points of the role
  概述職位的要點

- A clear outline of what is expected of your successor
  清楚敘述你對繼任者的期望

- A list of any essential files that will be handed over
  詳列將移交的任何重要文件

# 跨部門交接通知

To: juneblack@advertage.com
From: lylestone@advertage.com
Subject: Our New Graphic Designer

Hi June,

Not sure if you heard, but this is my last week at Advertage. It's been a great ride, but I've decided to take a position as art director in the firm GreenSky.

I've been training my replacement, Becky, for the last couple of weeks, and she's already **mastered**[1] our graphic-design style. She may be a little slow right now, but she's **bright**[2] and I'm sure she'll pick up her speed in no time. I hope you can give her some **feedback**[3] about her work when you see it.

> 指「立即；很快地」，意同 immediately、at once。

We'll have the design for the Layman Bros. ad done tomorrow, by the way, which means there's nothing else in the **queue**[4] for now.

You can always contact me at my personal e-mail address if needed: lyle79@gmail.com.

It's been a pleasure!

Take care,

Lyle

## 中譯

嗨，瓊恩：

不確定你是否聽說了，但這是我在優勢廣告公司的最後一個禮拜。這一路走來非常開心，但我已決定接下綠天公司的藝術總監一職。

前幾週我一直在訓練我的接替人員貝琪，而她已掌握我們的平面設計風格。她現在可能動作有點慢，但她很聰明，我相信她很快就會趕上速度。我希望你在看到她的設計時能給她一些意見。

對了，我們會在明天完成磊曼兄弟的廣告設計，這代表目前不會有其他東西要排著做。

如有需要，你可以用我的私人電子信箱 **lyle79@gmail.com** 跟我聯絡。

很榮幸能跟你一起共事！

保重，

萊爾

## 字彙

*1.* **master** [ˋmæstɚ] *v.* 掌握；精通

*2.* **bright** [braɪt] *adj.* 聰明的

*3.* **feedback** [ˋfidˏbæk] *n.* 反饋意見

*4.* **queue** [kju] *n.*（人事物的）行列

---

**參考答案**

1. Keep in mind that you should hand over the weekly reports every Monday morning.

2. Questions about travel allowances can be directed to Nancy in Accounting.

3. I've been training my replacement, John, for the last two weeks, and he's already mastered our logistics management.

4. I have made the necessary arrangements of handing over all my duties to my replacement, Sarah Black.

# Job
# Responsibilities
## 工作業務篇

簽呈

會議紀錄

跨部門業務
需求信

反映信

休假通知

出差行程
確認信

出差報告

出差後
謝函

業績報告

# Requests for Approval
## 簽呈

## 課前寫作練習

簽呈怎麼寫？
請參考主題詞彙、中譯及括號內的英文提示，將下列句子翻譯成英文。

❶ 來信是為了請您准許展開和一名潛在客戶的合作協議。

 ────────────────────────────

────────────────────────────

主題
詞彙

- **approval** [əˋpruvəl]
  批核
- **authorization**
  [ˌɔθərəˋzeʃən] 批准

- **comment** [ˋkɑˌmɛnt]
  意見
- **decision** [dɪˋsɪʒən]
  決定

❷ 以下是一份提案，籲請您批准在我們與星際公司
（**Star Corp.**）的協商中提供 **5%** 的折扣。

㊢ _____

_____

❸ 就我的專業意見來看，我們可輕而易舉地將服務拓
展至國際市場。

㊢ _____

_____

_____

❹ 感謝您能對增加網路花費的提案提出意見。

㊢ _____

_____

參考答案請見 p. 113

- **go-ahead** [ˋgoəˌhɛd]
  許可
- **grant** [grænt]
  准予

- **recommend** [ˌrɛkəˋmɛnd]
  建議
- **request** [rɪˋkwɛst]
  請求

TRACK 033

## 1 I'm writing to request the go-ahead to . . .

來信是為了請您准許……

- **I'm writing to request the go-ahead to purchase the additional machinery required for our new branch.**
  來信是為了請您准許添購新分公司所需的額外機器。

 相關寫法
- **The following is a contract from our new supplier** which **requires your authorization.**
  以下是我們新供應商的合約，需要您的批准。

## 2 I recommend we . . . for the following reasons:

我建議我們……，理由如下：

- **I recommend we expand employee health care coverage to part-time workers** for the following reasons:
  我建議我們將員工健保擴及兼職人員，理由如下：

 相關寫法
- I suggest that we **do not renew our exclusivity agreement with our current distributor** for the good of of the company.
  為了公司著想，我建議我們不與目前的經銷商更新獨家協議。

## 3 It is my professional opinion that . . .

就我的專業意見來看……

- It is my professional opinion that **transitioning to a paperless office would reduce operation costs.**
  就我的專業意見來看，轉為無紙化辦公室能減少營運成本。

- I am of the opinion that **investing in solar panels would pay off in the long term.**
  我認為投資太陽能板長期下來會有收穫。

- The approval of the policy would facilitate **an increase in demand from the younger demographic.**
  此項政策的批准可望使年輕族群的需求增加。

---

## 4 I would appreciate your comments regarding . . .

感謝您能對……提出意見。

- I would appreciate your comments regarding **my list of recommended policy modifications.**
  感謝您能對我建議的政策修訂清單提出意見。

- I look forward to hearing your thoughts and hope we can go ahead with **the proposed changes to the advertisement.**
  我期待得知您的想法，並希望我們能進行所提議的廣告變動。

- I would appreciate your response **in a timely manner.**
  感謝您能及時回覆。

# 徵詢主管意見

To: jbatra@sabreelectronics.com
From: dmalwalki@sabreelectronics.com
Subject: Request for Collaboration

Dear Mr. Batra,

> cross paths with sb 指「偶遇某人」。

I'm writing to request the go-ahead to follow up on an account. While at the New York Technology Trade Show, I crossed paths with a Benson Technologies representative who expressed interest in contracting Sabre Electronics as a supplier. I recommend we **pursue**[1] the collaboration for the following reasons:

- We could expect orders valued at over US$10 million annually.
- We would **secure**[2] a global enterprise as a long-term customer.
- We would be able to expand our presence along the Eastern Seaboard.

> 指「簡言之、概括地說」。

> 指「美國東岸」，或稱「大西洋海岸」。

However, the **drawbacks**[3] — in a nutshell — are that we'll have to **wade through**[4] the **bureaucracy**[5] of a supplier **audit**[6] and acquire extra workers and production lines, since **lead time**[7] is one of the client's major concerns.

It is my professional opinion that this contract represents a great opportunity to expand our operations. I would appreciate your comments regarding the situation and await your final decision.

Best,

Devon Malwalki

中譯

巴特拉先生您好，

来信是為了請您准許我和一名客戶進行後續的往來。在紐約科技商展時，我遇到一位班森科技公司的代表，他有意和賽博電子簽訂供應商合約。我建議我們追求這個合作機會，理由如下：

- 我們可預期每年的訂單價值超過一千萬美元。
- 我們將有全球性的企業做為公司的長期客戶。
- 我們可沿著美國東岸拓展公司的能見度。

然而，不利因素，簡言之，是我們必須費力完成供應商審核的繁冗作業，還得聘用額外的員工並加開產線，因為交期是該客戶的主要考量之一。

就我的專業意見來看，此合約代表的是拓展公司營運的大好機會。感謝您能對此情形提出意見，我靜候您的最終決定。

祝安好，

戴文・馬爾瓦基

## 字彙

1. **pursue** [pə`su] *v.* 追求
2. **secure** [sə`kjur] *v.* 獲得
3. **drawback** [`drɔ‚bæk] *n.* 缺點
4. **wade through** *v.* 費力地做
5. **bureaucracy** [bju`rɑkrəsi] *n.* 官僚體制
6. **audit** [`ɔdət] *n.* 審查
7. **lead time** [lid] *n.*（產品）交期

 延伸學習　「批核；同意」相關詞彙

- **allowance** [ə`lauəns]
- **consent** [kən`sɛnt]
- **endorsement** [ɪn`dɔrsmənt]
- **green light**
- **permission** [pə`mɪʃən]
- **thumbs-up** [`θʌmz‚ʌp]

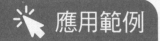 

# 向主管建議

| | |
|---|---|
| To: | mmalloy@taoscorp.com |
| From: | amuller@taoscorp.com |
| Subject: | Proposals to Reduce Company Expenses |
| Attached: | Price List.xlsx; Savings.xlsx |

Dear Ms. Malloy,

The following is a list of proposals to reduce company **overhead**[1] which requires your authorization.

• The printers at our locations are in need of replacement. I have attached the product price list of a suitable supplier and a chart showing our predicted annual savings per model for your consideration.

• I recommend that we sell our company cars and start a corporate account with Uber to transport sales representatives to visit clients.

• I suggest that we **curb**[2] business trip expenses for the good of the company. This can be achieved by restricting employees across the board from taking business class flights and reducing hotel **accommodation**[3] fees to US$130 per night.

指「全面性地；遍及各領域和階層」。

Thank you in advance for your **timely**[4] response. I look forward to hearing your thoughts and hope we can go ahead with the proposed changes.

Regards,

Alexei Muller

中譯

瑪洛伊女士您好，

以下是減少公司經常性開支的建議清單，需要您的批准。

- 營運據點的印表機需要更換。我已附上合適供應商的產品價目表以及顯示各機型每年預估可節省之經費的圖表，供您參考。
- 我建議賣掉公司車，並開設 **Uber** 的企業帳戶來載送業務代表去拜訪客戶。
- 為了公司著想，我建議我們控制差旅費用。這可藉由限制全體職員不得搭乘商務艙與降低飯店住宿費至每晚一百三十美元來達成。

預先感謝您的及時回覆。我期待得知您的想法，並希望我們能進行所提議的改變。

謹致，

阿列克謝・穆勒

## 字彙

1. **overhead** [ˋovɚˏhɛd]
   n.（公司）經常性開支

2. **curb** [kɝb] v. 限制

3. **accommodation** [əˏkɑməˋdeʃən]
   n. 住處

4. **timely** [ˋtaɪmlɪ] adj. 及時的

---

參考答案

1. I'm writing to request the go-ahead to pursue a partnership agreement with a prospective client.

2. The following is a proposal to request your approval to offer a 5 percent discount in our negotiations with Star Corp.

3. It is my professional opinion that we could easily expand our services to the international market.

4. I would appreciate your comments regarding the proposal to up online spending.

# Meeting Minutes
## 會議紀錄

## 課前寫作練習

會議紀錄怎麼寫？
請參考主題詞彙、中譯及括號內的英文提示，將下列句子翻譯成英文。

❶ 阿加莎．凱（Agatha Kay）於早上十點召開會議。

_____

_____

主題詞彙

- **absentee** [ˌæbsənˋti]
  缺席者
- **adjourn** [əˋdʒɝn]
  休會
- **agenda** [əˋdʒɛndə]
  議程
- **agree** [əˋgri]
  同意

❷ 恩立奎（Enrique）扼要陳述 6 月 5 日所舉行會議的重點。

(寫)
_____

_____

❸ 增加差旅津貼的提議以六比四的票數否決。

(寫)
_____

_____

❹ 全體一致同意公司應採用彈性工時策略。

(寫)
_____

_____

參考答案請見 **p. 121**

- **attendee** [ə͵tɛn`di]
  與會者
- **minutes** [`mɪnəts]
  會議紀錄
- **object** [əb`dʒɛkt]
  反對
- **propose** [prə`poz]
  提議

## 重點句型

**1** **(Name) called the meeting to order at (time).**

（人名）於（時間）召開會議。

- **Katharine Parks called the meeting to order at 9:30 a.m.**
  凱瑟琳・帕克斯於早上九點半召開會議。

> call to order 指「正式展開」。

**2** **(Name) ran/went over the highlights from the meeting held on (date).**

（人名）扼要陳述（日期）所舉行會議的重點。

- **Marcia went over the highlights from the meeting held on December 10.**
  瑪西亞扼要陳述 12 月 10 日所舉行會議的重點。

**3** **. . . was approved/rejected by a vote of (number) to (number).**

……以（數字）比（數字）的票數通過／否決。

- **The proposal to switch suppliers was approved by a vote of five to three.**
  更換供應商的提議以五比三的票數通過。

## 4 It was unanimously agreed that . . .

全體一致同意……

- **It was unanimously agreed that budget cuts made last year would remain in place.**
  全體一致同意去年制定的預算削減維持不變。

## 5 (Name) called for suggestions to . . .

（人名）徵求……的建議。

- **Darrel called for suggestions to improve the company's public image.**
  達瑞爾徵求改善公司公眾形象的建議。

## 6 (The) meeting (was) adjourned at (time).

會議於（時間）結束。

- **The meeting was adjourned at 4:15 p.m.**
  會議於下午四點十五分結束。

TRACK 037

# 一般會議紀錄

**Flora Lighting Inc.**
**New Products**

**Date & Time:** 3:30 p.m., March 15
**Location:** The executive conference room
**Present:** Cameron Mundy (Chair), Anderson Styles,
Bethany Wright, Laura Onassis, Arthur Coldwell,
Carly Norris (minutes taker)

Cameron Mundy called the meeting to order at 3:30 p.m.

The highlights of the meeting are as follows:

I.    Carly Norris ran over the highlights from the meeting held on February 27.

II.   Bethany Wright took the floor to present the new line of lights.

> floor 指「發言權」，take the floor 即「發言」之意。

III.  Arthur Coldwell objected to the copper fittings being used, **indicating**[1] copper-plated **nickel**[2] fittings would be less expensive.

IV.   It was unanimously agreed that the design team should **switch**[3] the **fittings**.[4]

V.    Laura Onassis proposed a **revamp**[5] of the company's kitchen light line.

VI.   Bethany Wright said such considerations were **underway**,[6] but it would take some time.

VII.  Anderson Styles proposed investing greater resources to redesign the line of kitchen lights.

VIII. Cameron Mundy objected on the grounds that it was not a priority.

IX.   It was agreed that Laura Onassis' proposal would be considered during the next meeting.

Meeting adjourned at 6 p.m.

Minutes submitted by Carly Norris.

## 中譯

花漾燈飾公司
新品會議

日期與時間：**3 月 15 日**下午三點半
地點：行政會議室
與會者：卡麥隆・蒙迪（主席）、安德森・
　　　史戴爾斯、貝瑟妮・萊特、蘿拉・
　　　歐納西斯、亞瑟・寇德威爾、卡
　　　莉・諾莉斯（會議記錄人）

卡麥隆・蒙迪於下午三點半召開會議。

會議重點如下：

一、 卡莉・諾莉斯扼要陳述 2 月 27 日所舉
　　 行會議的重點。

二、 貝瑟妮・萊特發言簡報新燈飾系列。

三、 亞瑟・寇德威爾反對使用中的銅材配
　　 飾，指出鍍銅的鎳材配飾會便宜些。

四、 全體一致同意設計團隊應更換配飾。

五、 蘿拉・歐納西斯提議改造公司的廚房
　　 燈具系列。

六、 貝瑟妮・萊特表示已在進行此考量，
　　 但會需要一些時間。

七、 安德森・史戴爾斯提議挹注更多資源
　　 於重新設計廚房燈具系列。

八、 卡麥隆・蒙迪反對的理由是此非首要
　　 之務。

九、 大家同意蘿拉・歐納西斯的提議會在
　　 下次的會議中予以斟酌。

會議於下午六點結束。

會議紀錄由卡莉・諾莉斯提交。

## 字彙

1. **indicate** [ˋɪndəˌket] *v.* 表示

2. **nickel** [ˋnɪkəl] *n.* 鎳

3. **switch** [swɪtʃ] *v.* 更換

4. **fitting** [ˋfɪtɪŋ] *n.* 配飾

5. **revamp** [riˋvæmp] *n.* 改造；翻新

6. **underway** [ˌʌndəˋwe]
 *adj.* 進行中的

### 延伸學習　會議紀錄 寫作要點

- **Date and time of the meeting**
 會議的日期和時間

- **Names of the participants and those unable to attend**
 與會者和無法參加者的姓名

- **Acceptance or amendments to previous meeting minutes**
 前次會議紀錄同意或修改的事項

- **Decisions made about each agenda item**
 每個議程項目的決定內容

# 正式會議紀錄

**Sequoia Designs
Senior Executive Meeting: Outlook for Q2**

**Date and location:** January 4 in the fourth-floor meeting room
**Attendees:** Johan Stevens (Chair), Hannah Barnes,
Toby Schwartz, Kirsten Chase, Yusuf Rushdie
**Absent with apologies:** Apologies were received from Jamal
Rice who was overseas.
**Minutes taken by:** Yusuf Rushdie
**Call to order:** 9 a.m. by Johan Stevens

**Meeting agenda:**
- Minutes from meeting held December 12
  - Minutes read by Yusuf, confirmed by Chair.

- Reports from departmental managers
  - Hannah (Sales): Q2 sales **projections**[1]
  - Kirsten (Marketing): advertising strategy for Q2
  - Toby (R & D): projects in the pipeline

指「在進行中；在籌畫中」。

- Goals
  - Johan called for suggestions to increase sales and cut costs.
  - Hannah **resolved**[2] to increase sales in three key regions.
  - Toby promised a shorter **turnover**[3] for new products.

- Proposals
  - Kirsten suggested spending more on online advertising.
  - Johan moved to put off the motion.
  - Kirsten's proposal was rejected by a vote of three to two.

**Meeting adjourned** at 12 p.m.

**Next meeting:** The **committee**[4] agreed that the next meeting
will be held on February 22.

## 中譯

### 紅杉設計公司
### 高階主管會議：第二季展望

**日期與地點：**1 月 4 日假四樓會議室

**與會者：**行政會議室

**與會者：**約翰·史蒂文斯（主席）、漢娜·
巴恩斯、托比·施華茲、克絲汀·
查斯、尤瑟夫·魯西迪

**缺席致歉者：**已收到人在國外的賈瑪爾·萊
斯的致歉

**會議記錄人：**尤瑟夫·魯西迪

**會議召開：**約翰·史蒂文斯於早上九點召開

**議程：**

• **12 月 12 日的會議紀錄**

▪ 尤瑟夫宣讀會議紀錄，主席予以確認。

• **各部門經理會報**

▪ 漢娜（業務部）：第二季銷售預估

▪ 克絲汀（行銷部）：第二季廣告策略

▪ 托比（研發部）：進行中的專案

• **目標**

▪ 約翰徵求增加銷量與減少成本的建議。

▪ 漢娜決意增加三個重點區域的銷量。

▪ 托比承諾縮短新產品的周轉期。

• **提案**

▪ 克絲汀建議增加網路廣告的支出。

▪ 約翰提議延緩此動議。

▪ 克絲汀的提議以三比二的票數否決。

**休會**於中午十二點。

**下次會議：**委員會同意下一場會議於 2 月
22 日舉行。

## 字彙

1. **projection** [prəˋdʒɛkʃən] *n.* 預估

2. **resolve** [rɪˋzɑlv] *v.* 決意

3. **turnover** [ˋtɜn͵ovɚ] *n.* 周轉期

4. **committee** [kəˋmɪtɪ] *n.* 委員會

參考答案

1. Agatha Kay called the meeting to order at 10 a.m.

2. Enrique ran/went over the highlights from the meeting held on June 5.

3. The proposal to increase travel allowance was rejected by a vote of six to four.

4. It was unanimously agreed that the company should adopt the flextime policy.

# Unit 13

# Cross-Departmental Help Requests

## 跨部門業務需求信

### 課前寫作練習

跨部門業務需求信怎麼寫？
請參考主題詞彙、中譯及括號內的英文提示，將下列句子翻譯成英文。

**❶** 我們將需要幾名財務部人員支援兩天。

_____

_____

主題
詞彙

- **assist** [əˋsɪst]
  協助
- **available** [əˋveləbəl]
  有空的

- **coordinate** [koˋɔrdəˏnet]
  協調
- **handle** [hændḷ]
  處理

❷ 你能讓我知道誰有空支援校稿嗎？

（寫）

＿＿＿＿＿＿＿＿＿＿＿＿＿＿＿＿＿＿＿＿

＿＿＿＿＿＿＿＿＿＿＿＿＿＿＿＿＿＿＿＿

❸ 這張海報須重做，這已超出我的平面設計師可處理的量了。

（寫）

＿＿＿＿＿＿＿＿＿＿＿＿＿＿＿＿＿＿＿＿

＿＿＿＿＿＿＿＿＿＿＿＿＿＿＿＿＿＿＿＿

❹ 你可能調安娜（Anna）來協助我們翻譯嗎？

（寫）

＿＿＿＿＿＿＿＿＿＿＿＿＿＿＿＿＿＿＿＿

＿＿＿＿＿＿＿＿＿＿＿＿＿＿＿＿＿＿＿＿

參考答案請見 p. 129

- **prioritize** [praɪˋɔrəˌtaɪz]
  確定優先次序
- **requirement** [rɪˋkwaɪrmənt]
  需求
- **spare** [spɛr]
  騰出（人力、時間）
- **squeeze in** [skwiz]
  擠出時間做

## 1 We'll need several (department) people for (period of time).

我們將需要幾名（部門名稱）人員支援（一段時間）。

- **We'll need several Art Design people for two weeks.**
  我們將需要幾名藝術設計人員支援兩週。

 相關寫法
- Which day this week can I book someone to **set up my team's computers at the new location?**
  這星期哪天我可以預約人員在新據點架設我小組成員的電腦？

---

## 2 Can you let me know who will be available for (type of task)?

你能讓我知道誰有空支援（工作類別）嗎？

- Can you let me know who will be available for **the routine maintenance?**
  你能讓我知道誰有空支援例行維修的作業嗎？

 相關寫法
- I need **an IT specialist to join** a software review panel **to ensure the final choice is compatible with our servers.**
  我需要一名資訊科技專員加入軟體評鑑小組，確保最終的選擇能和我們的伺服器相容。

## 3 The . . . needs to be redone, which is simply more than (name) can handle alone.

……須重做,這已超出(人名)可處理的量了。

- The **banner ad** needs to be redone, which is simply more than **my design team** can handle alone.

  這個橫幅廣告須重做,這已超出我的設計團隊可處理的量了。

- Would **Rebecca** be available for **a short translation job** that shouldn't take more than **two hours**?

  蕾貝卡有空協助一個應該不會超過兩小時的簡短翻譯工作嗎?

---

## 4 Could you possibly spare (name) to help us out with . . .?

你可能調(人名)來協助我們……嗎?

- Could you possibly spare **Tim** to help us out with **copyediting**?

  你可能調提姆來協助我們審稿嗎?

- Could someone on your team spare **an hour** to **advise me on writing a contract**?

  你小組裡有誰可以撥一小時來指導我寫合約呢?

- The next time **Glen** has a **half-hour** window, I would very much appreciate if **he could help us out**.

  下次葛蘭有半小時的空檔時,如果他能幫我們忙的話,我會非常感激。

 TRACK 040

# 技術支援篇

To: dhowell@webdez.com
From: iwright@webdez.com
Subject: Requesting IT Support

Hello, Debra:

How are you?

As we **briefly**[1] discussed in the managers' meeting on Tuesday, we just won a **bid**[2] for an important contract with Becker Investments. We will be designing an online shopping mall that has already signed on 60 brands. This will be **similar**[3] to the French project last year, where we'll need several **dedicated**[4] IT people for one to two months once the testing begins in mid-April. Can you let me know who will be available for this? I'd like to keep them in the loop from the beginning.

> keep sb in the loop 指「隨時讓某人知道進展」。

Many thanks,

Ivan Wright
Project Manager
WebDez, Inc.

## 中譯

哈囉，黛柏拉：

你好嗎？

如同我們在星期二的主管會議上簡短討論到的，我們剛競標贏得貝克投資公司的重要合約。我們將設計一個已和六十個品牌簽約的網路商城。這和去年的法國專案會很相似，一旦測試階段於四月中開始，我們將需要幾名專職的資訊技術人員支援一到兩個月。你能讓我知道誰有空支援嗎？我想讓他們一開始就知道所有的進展。

非常感謝，

艾文・萊特
專案經理
戴斯網路公司

## 字彙

1. **briefly** [ˋbrifli] *adv.* 簡短地

2. **bid** [bɪd] *n.* 競價；投標

3. **similar** [ˋsɪmələ] *adj.* 相似的（+ to）

4. **dedicated** [ˋdɛdɪͺketəd] *adj.* 專職的

## 延伸學習 　跨部門需求信 寫作要點

- Use a proper greeting.
  使用恰當的問候語。

- Describe your issue in detail.
  詳述你遇到的問題。

- State the favor you're asking.
  陳述你需要的協助。

- Tell your recipient why the favor is important.
  告知收信者此協助的重要性。

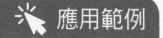

# 人力支援篇

To: rkumer@inroadspublishing.com
From: bwalters@inroadspublishing.com
Subject: Requesting Design Support

Good morning, Ravi.

Here are the details you requested. We've just been given a **rush**[1] contract for our *Science Explorers* series to be published in Italian. The first seven books have already been translated, and there are nine in total.

> 於此為介系詞，指「並且、而且；加上」。

The entire **layout**[2] needs to be **redone**,[3] which is simply more than my designer can handle alone, plus he needs to create new front and back covers. Could you possibly spare Alison to help us out with the layout? Of course, she would **document**[4] her time and I'll be happy to **reciprocate**[5] in the future.

Sincerely,

Brittany Walters
Executive Editor
InRoads Publishing

## 中譯

拉維，早安。

這封信有你之前問到的細節。我們剛拿到一筆很趕的合約，要以義大利文出版公司的《科學探險》系列叢書。前七冊已翻譯好，總共有九本。

此系列的整個版面須重做，這已超出我的美編可處理的量了，而且他還得設計新的封面和封底。你可能調艾莉森來協助我們排版嗎？當然，她可以記下她花的時間，日後我會很樂意回報的。

謹啟，

布蘭妮・瓦特斯
執行編輯
印路出版社

## 字彙

1. **rush** [rʌʃ] *adj.* 緊急的
2. **layout** [ˋleˏaʊt] *n.* 版面設計
3. **redo** [riˋdu] *v.* 重做
4. **document** [ˋdɑkjəˏmɛnt] *v.* 記錄
5. **reciprocate** [rɪˋsɪprəˏket] *v.* 回報

參考答案

1. We'll need several Finance people for two days.

2. Can you let me know who will be available for proofreading?

3. The poster needs to be redone, which is simply more than my graphic designer can handle alone.

4. Could you possibly spare Anna to help us out with the translation job?

# Opinion Letters
# 反映信

## 課前寫作練習

反映信怎麼寫？
請參考主題詞彙和中譯，將下列句子翻譯成英文。

❶ 我想投訴多功能事務機的機械故障。

 ......................................................................................

......................................................................................

主題
詞彙

- **address** [əˋdrɛs]
  處理
- **complaint** [kəmˋplent]
  投訴；不滿
- **consideration** [kənˌsɪdəˋreʃən]
  體諒；考慮
- **disturb** [dɪˋstɝb]
  打擾

❷ 我很確定其他人也同樣因為不斷的高分貝宣布聲而感到不便，希望這些干擾能就此打住。

寫 _____

_____

❸ 這名接待員對於任何在午休時間進來辦公室的人都會整個反應過度。

寫 _____

_____

❹ 在他下星期去度假以前若能解決分歧的意見就太好了。

寫 _____

參考答案請見 p. 137

- **inconvenience**
  [ˌɪnkənˋvinjəns] 帶來不便

- **lodge** [lɑdʒ]
  提出（申訴、抗議）

- **resolve** [rɪˋzɑlv]
  解決

- **unacceptable**
  [ˌʌnɪkˋsɛptəbəl] 不能接受的

## 1 I would like to lodge a complaint about . . .

我想投訴……

- I would like to lodge a complaint about **the regular use of the copy machine during the quiet lunch break.**

  我想投訴安靜午休時常會有人使用影印機。

 相關寫法
- I am unhappy to report that **one of the toilet seats in the restroom is cracked.**

  我很不開心要舉報洗手間的其中一個馬桶座圈裂掉了。

---

## 2 I am quite certain that others are just as inconvenienced by . . .

我很確定其他人也同樣因為……而感到不便。

- I am quite certain that others are just as inconvenienced by **the garbage that's simply left in front of the building's door.**

  我很確定其他人也同樣因為放在大樓門口前面的垃圾而感到不便。

 相關寫法
- I am most disturbed by regularly finding **my mail forcibly shoved into the mailbox even when it clearly doesn't fit.**

  最讓我困擾的是我常常發現我的郵件即使很明顯放不進去，還是會被強行塞進郵箱裡。

## 3 Basically, we are all experiencing major friction with (name).

基本上，我們都和（人名）出現極大的摩擦。

- **Basically, we are all experiencing major friction with the new volunteer coordinator.**

  基本上，我們都和新的志工協調員出現極大的摩擦。

 • Angela completely overreacts to any request for additional support from her team.

安潔拉對於任何要求其小組應給予的額外支援都會整個反應過度。

---

## 4 I hope (department) can address . . .

我希望（部門名）能處理……

- **I hope Accounts Payable can address the outstanding sum within two weeks; otherwise, we will stop regular deliveries.**

  我希望應付帳款部門能在兩星期內處理未付的帳款總額，不然我們將停止正常運送。

 • It would be great to resolve the client's account problems in a timely manner.

如果能盡快解決客戶的帳戶問題就太好了。

- I believe there must be standard operating procedures regarding this type of complaint?

我相信有關這類的投訴一定有標準作業程序吧？

# 投訴公德心問題

To: gdavis@newlife.com
From: ewright@newlife.com
Subject: Behavioral Complaints

Dear Ms. Davis,

I would like to lodge a complaint about certain **ongoing**[1] behaviors at the office that I find increasingly unacceptable.

In my view, they all **stem from**[2] lack of consideration for others. Undoubtedly, there is no **malicious**[3] **intent**,[4] but I am quite certain that others are just as inconvenienced by these problems.

I am most disturbed by regularly finding a wet seat when I use the restroom, as it is simply **unhygienic!**[5] I've also noticed the regular use of the copy machine and shredder during what's meant to be a quiet hour at lunch, and the water dispenser is frequently **clogged**[6] with tea leaves or food **residue**.[7]

I hope the office administration can address these issues in a timely manner. Perhaps **prominent**[8] signs and a company-wide e-mail would do the trick?

> 指「盡快」，可用 in a timely fashion 替換。

> 指「起作用；奏效」，類似用法為 work wonders。

Many thanks,

Eric Wright

## 中譯

戴維斯女士您好，

我想投訴某些在辦公室裡持續發生，而且我越來越不能接受的的行為。

在我看來，這些行為都是因為沒有考慮他人而造成的。毫無疑問地，這不是出於惡意，但我很確定其他人也同樣因為這些問題而感到不便。

最讓我困擾的是我在使用洗手間時常常發現馬桶座椅溼溼的，整個就很不衛生！我也注意到本該是安靜的午餐時刻，卻常有人使用影印機和碎紙機，還有飲水機常因茶葉和食物殘渣而阻塞。

我希望管理部能盡快處理這些問題。或許顯眼的標示和寄信給全公司會有效？

非常感謝，

艾瑞克・萊特

## 字彙

1. **ongoing** [ˋɑnˏgoɪŋ] *adj.* 不間斷的

2. **stem from** [stɛm] *v.* 由……造成

3. **malicious** [məˋlɪʃəs] *adj.* 惡意的

4. **intent** [ɪnˋtɛnt] *n.* 目的；意圖

5. **unhygienic** [ˏʌnˏhaɪˋdʒɛnɪk] *adj.* 不衛生的

6. **clog** [klɑg] *v.* 阻塞

7. **residue** [ˋrɛzəˏdju] *n.* 殘餘物

8. **prominent** [ˋprɑmənənt] *adj.* 顯眼的

### 延伸學習　反應信 寫作要點

- **State the facts briefly, precisely, and clearly.**
  簡要、準確且清楚地陳述事實。

- **Explain how you feel about the behavior you are criticizing.**
  解釋你對於所批評行為的感受。

- **Avoid rudeness.**
  避免無禮。

- **Suggest desired results/action.**
  建議希望的結果／行動。

# 投訴個人行為問題

To: jim@nbccompany.com
From: anderson@nbccompany.com
Subject: Behavioral Complaints

Hi Jim,

I'll warn you right away. This is quite a serious e-mail.

Basically, we are all experiencing major friction with Nate. He completely overreacts to any kind of disagreement, including **constructive**[1] criticism. It's one thing to get **defensive**,[2] but Nate tends to explode in a **stream**[3] of ugly words that **border**[4] on **verbal abuse**.[5]

Yesterday afternoon in our weekly team meeting, Kate simply mentioned she'd found a **miscalculation**[6] in his report. He **literally**[7] shouted at her and stormed out of the room, leaving her in tears. He apologized this morning, but this is hardly an isolated incident. We're constantly walking on eggshells around him.

> storm out 指「生氣地跑出去」。

> walk on eggshells 指「小心謹慎」。

It would be great to resolve this issue **promptly**.[8] I believe there must be a company policy regarding this?

Thanks for your help,

Anderson

## 中譯

嗨，吉姆，

我要馬上提醒你。這是一封相當重要的電子郵件。

基本上，我們都和奈特出現極大的摩擦。他對於任何的意見不合都會整個反應過度，包含有建設性的批評在內。有防禦心是一回事，但奈特會爆出一連串幾近言語羞辱的難聽字眼。

昨天下午在每週的小組會議裡，凱特只是提到她在他的報告裡找到一個算錯的地方。他直接就吼她，然後衝出會議室，害她哭了出來。他今天早上道歉了，但這絕不是個案。我們跟他相處常常得戰戰兢兢的。

如果能立即解決這個問題就太好了。我相信關於這種問題一定有相應的公司政策吧？

感謝你的協助，

安德森

## 字彙

1. **constructive** [kən`strʌktɪv]
   *adj.* 有建設性的

2. **defensive** [dɪ`fɛnsɪv] *adj.* 防禦的

3. **stream** [strim] *n.* 一連串

4. **border** [`bɔrdə] *v.* 幾近（+ on）

5. **verbal abuse** [`vɜbəl] [ə`bjus]
   *n.* 言語羞辱

6. **miscalculation** [ˌmɪsˌkælkjə`leʃən]
   *n.* 算錯

7. **literally** [`lɪtərəli] *adv.* 直接地

8. **promptly** [`prɑmptli] *adv.* 立即地

---

參考答案

1. I would like to lodge a complaint about a technical malfunction with the multifunction printer.

2. I am quite certain that others are just as inconvenienced by the constant loud announcements and would be happy to put an end to the interruptions.

3. The receptionist completely overreacts to anyone coming into the office during the lunch break.

4. It would be great to resolve the disagreement before he goes on vacation next week.

# Vacation Notifications
## 休假通知

### 課前寫作練習

休假通知怎麼寫？
請參考主題詞彙、中譯及括號內的英文提示，將下列句子翻譯成英文。

❶ 此信旨在通知你，我將從 2 月 8 日開始休假，2 月 19 日回公司。

---

主題詞彙

- **absence** [ˋæbsəns]
  缺勤
- **deputy** [ˋdɛpjəti]
  職務代理人

- **in charge of** [tʃɑrdʒ]
  負責……
- **in sb's stead** [stɛd]
  代替某人

**❷** 我不在時，所有和假期促銷相關的事項將由我的職務代理人萊諾‧凱斯（Lionel Case）處理。

(寫) _____

_____

**❸** 暫代我職務的貝絲（Beth）會負責安德森客戶（Anderson account）。

(寫) _____

_____

**❹** 如有緊急情況發生，請與公關經理賽門（Simon）聯繫。

(寫) _____

_____

參考答案請見 p. 145

- **in touch**
  聯繫
- **on leave/vacation**
  休假

- **pending** [ˈpɛndɪŋ]
  待定的；未決的
- **urgent** [ˈɝdʒənt]
  急迫的

**1** **This is to inform you that I will be on leave/ vacation from (date), returning to the office on (date).**

此信旨在通知你，我將從（日期）開始休假，（日期）回公司。

• This is to inform you that I will be on vacation from **January 1**, returning to the office on **January 11**.

  此信旨在通知你，我將從 1 月 1 日開始休假，1 月 11 日回公司。

 • I will be on leave from **December 15** to **December 21**.

  我將從 12 月 15 日休到 12 月 21 日。

**2** **In my absence, all matters related to (work) will be handled by my deputy, (name).**

我不在時，所有和（工作）相關的事項將由我的職務代理人（人名）處理。

• In my absence, all matters related to **the ad campaign will be handled by my deputy, Ludwig Freeman**.

  我不在時，所有和廣告活動相關的事項將由我的職務代理人路德維希 · 費里曼處理。

 • In my stead, **Carla** will be in charge of **managing the project**.

  暫代我職務的卡拉將負責管理專案。

  • During my absence, **Luke Gregg** will be overseeing my desk.

  我不在的期間，路克 · 葛瑞格將負責我的工作。

# 3 A finalized (work) for . . . is due (date).

……的最終（工作），截止日是（日期）。

- **A finalized report for Mr. Thomson is due December 12.**
  要給唐姆森先生的最終報告，截止日是 12 月 12 日。

- **The budget for the Christmas party is ready for approval.**
  聖誕派對的預算已準備好送審了。

- **Still pending is the matter of confirming all the bookings in writing.**
  待定事項是書面確認所有的預訂項目。

---

# 4 If something urgent does come up, please get in touch with (name), (title).

如有緊急情況發生，請與（職稱）（人名）聯繫。

- **If something urgent does come up, please get in touch with Magda Richardson, the head of HR.**
  如有緊急情況發生，請與人資主管瑪格達‧理查森聯繫。

- **If you have any questions or concerns during this period, please contact Beverly Foxx at 555-4186.**
  若您在這段期間有任何問題或疑慮，請撥 555-4186 聯絡比佛莉‧福克斯。

- **You can contact Justin Brooke at extension 218 should you have any concerns before I return.**
  若您在我回來之前有任何不放心之處，可撥分機 218 與賈斯汀‧布魯克聯繫。

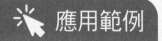 

# 部門工作安排

To: all@aionmedia.com

From: ltrang@aionmedia.com

Subject: Vacation Notice

Hi,

As you all well know, I will be on vacation from November 19 to December 12. In my stead, Robert will be in charge of **delegating**,[1] **supervising**,[2] and managing all ongoing work. I have full confidence that he and the team will rise to the occasion.

> 指「成功應對困難；臨危不亂」，
> occasion 可用 challenge 替換。

A few important things to keep in mind while I am away:

- A **finalized**[3] social media campaign for our news channel is due November 25.
- The second phase of **archiving**[4] our press releases needs to be completed by December 1.
- Work with the events and programming departments on our New Year's Eve broadcast is ongoing and should be **prioritized**.[5]

If something urgent does come up that Robert's not able to deal with, please get in touch with Mark Thompson, our senior manager at headquarters.

Best,

Luis Trang
Junior Public Relations Manager
Aion Media Group

## 中譯

嗨，

如你們所知，我將從 11 月 19 日休到 12 月 12 日。暫代我職務的羅伯特會負責分派、監督和管理所有正在進行的工作。我有十足的信心，他和整個團隊能臨危不亂。

我不在時，幾項重要的事情得記住：

· 新聞頻道的社群媒體活動，定案截止日是 11 月 25 日。

· 第二階段的新聞稿歸檔必須在 12 月 1 日前完成。

· 與活動規劃部合作的跨年夜廣播工作正在進行中，
  應優先處理。

如有羅伯特無法處理的緊急情況發生，請與總公司的
資深經理馬克‧湯普森聯繫。

祝好，

路易斯‧莊
公關襄理
艾昂媒體集團

## 字彙

1. **delegate** [ˋdɛləˏget] v. 委派

2. **supervise** [ˋsupɚˏvaɪz] v. 監督

3. **finalize** [ˋfaɪnəˏlaɪz] v. 完成

4. **archive** [ˋɑrˏkaɪv] v. 整理歸檔

5. **prioritize** [praɪˋɔrəˏtaɪz] v. 優先考慮

---

### 延伸學習　休假通知 必備撰寫事項

- **Dates of your absence**
  休假的日期

- **Name and contact info of your deputy**
  職務代理人的姓名和聯絡資訊

- **Delegated jobs and tasks in your absence**
  不在時所委任的工作和任務

- **Personal contact info for urgency**
  緊急情況的個人聯絡資訊

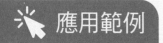

# 跨部門休假通知

To: ptravis@aionmedia.com
From: ltrang@aionmedia.com
Subject: Vacation Notice
Attached: Press Release.docx

Dear Travis,

This is to inform you that I will be on leave from November 19, returning to the office on December 12. In my absence, all matters related to our New Year's Eve **broadcast**[1] will be handled by my deputy, Robert Noah.

Please note the following tasks which were completed for the broadcast:

• Fan clubs for talent performing were contacted.
• Promotional posters were **approved**[2] and printed.
• A draft press release for the event was prepared and is ready for review (see attached).

Still pending is the matter of **coordinating**[3] with our print and radio partners to **implement**[4] our promotional strategy.

As far as our work on events for the coming year, unless urgent, they will have to be put on hold for now.

Best,

Luis Trang
Junior Public Relations Manager
Aion Media Group

put sth on hold 指「暫緩處理某事物」。

## 中譯

親愛的崔維斯，

此信旨在通知你，我將從 **11** 月 **19** 日開始休假，**12** 月 **12** 日回公司。我不在時，所有和跨年夜廣播相關的事項將由我的職務代理人羅伯特‧諾亞處理。

請留意以下和廣播有關的已完成項目：

· 才藝表演的粉絲團已取得聯繫。

· 宣傳海報已批准並印製。

· 活動的新聞稿草稿已備妥並準備進行複審（見附件）。

待定事項是與印刷廠和電台的合作同仁協調以執行我們的宣傳策略。

至於我們來年活動的工作，除非很緊急，否則目前都得暫緩處理。

祝好，

路易斯‧莊
公關襄理
艾昂媒體集團

## 字彙

*1.* **broadcast** [ˈbrɔdˌkæst] *n.* 廣播

*2.* **approve** [əˈpruv] *v.* 批准

*3.* **coordinate** [koˈɔrdəˌnet] *v.* 協調

*4.* **implement** [ˈɪmpləˌmɛnt] *v.* 執行

---

參考答案

1. This is to inform you that I will be on leave/vacation from February 8, returning to the office on February 19.

2. In my absence, all matters related to the holiday promotion will be handled by my deputy, Lionel Case.

3. In my stead, Beth will be in charge of the Anderson account.

4. If something urgent does come up, please get in touch with Simon, the PR manager.

# Business Trip Arrangements
## 出差行程確認信

---

## 課前寫作練習

出差行程確認信怎麼寫？
請參考主題詞彙、中譯及括號內的英文提示，將下列句子翻譯成英文。

❶ 謹代表薇拉・羅曼（**Willa Loman**），來信確認她
從台灣台北到西班牙馬德里的旅程安排。

---

**主題詞彙**

- **arrive** [əˋraɪv]
  抵達
- **confirm/verify**
  [kənˋfɝm] [ˋvɛrəˌfaɪ] 確認

- **depart** [dɪˋpɑrt]
  出發；啟程
- **duration** [dʊˋreʃən]
  持續期間

❷ 許女士預計於 9 月 12 日晚間 8 點從甘迺迪機場（JFK Airport）出發，並於當地時間下午 12 點抵達。

(寫) _____

_____

❸ 我想提醒您，詹斯女士（Ms. Janz）希望您能和她在歐迪恩餐廳（Odeon Restaurant）共進晚餐。

(寫) _____

_____

❹ 我想問問看是否有可能幫我們買到皇家馬德里（Real Madrid）足球賽的門票。

(寫) _____

_____

參考答案請見 p. 153

- **gratitude** [ˋgrætəˏtud]
  感謝
- **itinerary** [aɪˋtɪnəˏrɛri]
  預定行程
- **remind** [rɪˋmaɪnd]
  提醒
- **request** [rɪˋkwɛst]
  要求

# 1 On behalf of (name), I am writing to confirm (sb's) travel arrangements from (location) to (location).

謹代表（人名），來信確認他／她從（地點）到（地點）的旅程安排。

- **On behalf of Jenna Thomson, I am writing to confirm her travel arrangements from Taipei, Taiwan to Toronto, Canada.**

  謹代表珍娜・湯森，來信確認她從台灣台北到加拿大多倫多的旅程安排。

 • I am writing to confirm my travel details to **London for the annual Technology Exhibition in the city.**

來信確認我前往倫敦年度科技展的出差細節。

---

# 2 (Name) is set to depart from (location) on (date) at (time) and arrive at (time).

（人名）預計於（日期）（時間）從（地點）出發，並於（時間）抵達。

- **Ms. Janson is set to depart from Taiwan Taoyuan International Airport on June 25 at 8 p.m. and arrive at 10:45 p.m. local time.**

  詹森女士預計於 6 月 25 日晚間 8 點從台灣桃園國際機場出發，並於當地時間晚上 10 點 45 分抵達。

 • We will be arriving on **Iberia Airlines Flight 229 at 9:25 p.m. on Tuesday, December 14.**

我們將搭乘伊比利亞航空 229 航班於 12 月 14 日週二晚間 9 點 25 分抵達。

## 3 I would like to remind you . . .

我想提醒您……

- **I would like to remind you** of Ms. Downey's wish to have a window seat on both flights to and from Paris.
  我想提醒您,道尼女士希望往返巴黎的航班都坐靠窗的座位。

- We'd request that someone be at the airport to take **Ms. Vasquez** to **her hotel.**
  我們要求派人在機場接瓦斯奎茲女士到她下榻的飯店。
- **Mr. Street** is booked for a **two-night** stay at **the Royal Hotel.**
  史崔特先生預定在皇家飯店下榻兩晚。

---

## 4 I meant to ask if it was possible to . . .

我想問問看是否有可能……

- **I meant to ask if it was possible to** upgrade to first class on my return journey.
  我想問問看是否有可能將我的回程班機升等至頭等艙。

- We were wondering if it might be possible to **obtain tickets to a show at the Prague State Opera.**
  我們在想是否有可能買到布拉格國家歌劇院的演出門票。
- Could you please confirm if **there are conference rooms at the hotel?**
  麻煩您確認飯店內是否有會議室好嗎?

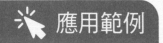 

# 代理確認函

To: bdoherty@polytech.com
From: psimons@veritastech.com
Subject: Travel Arrangements for Mr. Daniel Wilson

Dear Mr. Doherty,

On behalf of Mr. Daniel Wilson, I am writing to confirm his travel arrangements from Melbourne, Australia to Munich, Germany. According to our previous **correspondence**,[1] Mr. Wilson is set to depart from Melbourne Airport on May 29 at 8 a.m. and arrive at 1:20 p.m. the next day **inclusive**[2] of a short **layover**[3] in Dubai. I would also like to remind you of Ozzie Travel's promised round-trip upgrade to business class.

> 副詞，指「此外；同時」，替換用法
> 有 in addition、furthermore。

**Additionally**, as Mr. Wilson will be traveling between cities, please have an English-speaking driver and a car ready for the duration of the trip. We'd also request that you double-check the driver has a clean driving record.

We **extend**[4] our sincere gratitude for your much appreciated assistance on the matter.

Best regards,

Pauline Simons
Executive Assistant
Veritas Tech

## 中譯

道荷蒂先生您好，

謹代表丹尼爾‧威爾森先生，來信確認他從澳洲墨爾本到德國慕尼黑的旅程安排。根據我們先前的通信內容，威爾森先生預計於 5 月 29 日早上 8 點從墨爾本機場出發，並於隔天下午 1 點 20 分抵達，當中包含在杜拜的短暫停留。我也想提醒您，歐茲旅行社承諾會將來回機票升等至商務艙。

此外，由於威爾森先生將於各城市間來回行動，請在旅程期間安排好一名會說英語的司機和一輛用車。我們也要求您應加以確認該司機有良好的駕駛紀錄。

我們竭誠感謝您對此事的鼎力協助。

謹致，

寶琳‧西蒙斯
特別助理
維利塔斯科技

## 字彙

1. **correspondence** [ˌkɔrəˈspɑndəns]
   *n.* 通信

2. **inclusive** [ɪnˈklusɪv]
   *adj.* 包含的（+ of）

3. **layover** [ˈleˌovə]
   *n.* （飛行途中的）停留

4. **extend** [ɪkˈstɛnd] *v.* 給予

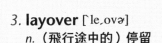

延伸
學習　**出差確認信 寫作要點**

- **trip purpose**
  出差目的

- **flight information**
  航班資訊

- **length of stay**
  停留時間

- **accommodation details**
  住宿細節

- **extra needs**
  額外需求

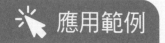

# 自行確認函

To: irene@mediworld.org
From: jupdike@rma.org
Subject: Travel Plans for James Updike

Dear Irene,

This is to verify that my travel plans have been finalized. Could you please confirm the following?

1. I depart Dallas Airport on May 15 at 9 a.m.

2. The return flight from Miami International leaves on the 18th at 8 p.m.

3. Jessica Tucker, the program **coordinator**,[1] will **receive**[2] me at the airport.

4. I'm booked at the M-Hotel for the duration of my stay.

Additionally, I meant to ask if it was possible to have a half-day city tour scheduled for my last day. It also **occurred to**[3] me to inform you that I am a **vegetarian**.[4]

Lastly, I just want to make sure everything is in order for my 6 p.m. **keynote**[5] speech on the 17th at the **convention**[6] center. Of course, I'm delighted to have this **outstanding**[7] opportunity and look forward to hearing from you soon.

Sincerely,

Dr. James Updike
Dean of Faculty of Medicine
Robarts Medical Academy

## 中譯

艾琳你好，

來信旨在確認我的旅程計畫已定案。麻煩你確認以下事項好嗎？

1. 我在 5 月 15 日早上 9 點從達拉斯機場出發。
2. 從邁阿密國際機場的回程班機於 18 日晚間 8 點起飛。
3. 節目統籌潔西卡‧塔克會在機場接我。
4. 停留期間我預定下榻 M 飯店。

此外，我想問問是否有可能在我行程的最後一天，安排一個半日的城市導覽。我也想到要告知你我吃素。

最後，我只是要確認 17 日晚間 6 點我在會議中心演講的所有事項都已安排妥當。當然，我很高興有這個寶貴的機會，並期待很快收到你的回覆。

謹啟，

詹姆斯‧厄普戴克博士
藥學院院長
羅巴茲醫療學院

## 字彙

1. **coordinator** [koˋɔrdəˌnetɚ] *n.* 統籌者

2. **receive** [rɪˋsiv] *v.* 迎接；歡迎

3. **occur to** [əˋkɝ] *v.* 想到

4. **vegetarian** [ˌvɛdʒəˋtɛrɪən] *n.* 素食者

5. **keynote** [ˋkiˌnot] *n.* 主講

6. **convention** [kənˋvɛnʃən] *n.* 會議

7. **outstanding** [autˋstændɪŋ] *adj.* 重要的

---

**參考答案**

1. On behalf of Willa Loman, I am writing to confirm her travel arrangements from Taipei, Taiwan to Madrid, Spain.

2. Ms. Hsu is set to depart from JFK Airport on September 12 at 8 p.m. and arrive at 12 p.m. local time.

3. I would like to remind you that Ms. Janz hopes you will join her for dinner at the Odeon Restaurant.

4. I meant to ask if it was possible to get us tickets to a Real Madrid soccer match.

# Business Trip Reports
## 出差報告

## 課前寫作練習

出差報告怎麼寫？
請參考主題詞彙、中譯及括號內的英文提示，將下列句子翻譯成英文。

❶ 我與業務總監哈利（**Harry**）會面，並向他介紹我們的資深工程師大衛（**David**）。

主題
詞彙

- **breakdown** [`brek͵daʊn]
  細項
- **conclusion** [kən`kluʒən]
  結論

- **event** [ɪ`vɛnt]
  事件
- **follow-up** [`falo͵ʌp]
  後續（行動）

❷ 我們評估問題在於長時間暴露在直射的陽光下。

(寫) _____

_____

❸ 初步結果顯示問題出在過時的作業系統。

(寫) _____

_____

❹ 我們提議進一步研究以找出真正的問題。

(寫) _____

_____

參考答案請見 p. 161

- **negotiation** [nɪˌgoʃiˋeʃən]
  協商
- **overview** [ˋovɚ͵vju]
  概述

- **submit** [səbˋmɪt]
  提交
- **summary** [ˋsʌməri]
  摘要

## 1 I met with (name), (title), and introduced him/her to our (title), (name).

我與（職稱）（人名）會面，並向他／她介紹我們的（職稱）（人名）。

- **I met with Harold, the head sales representative,** and introduced him to our **site technician, Ian.**
  我與首席業務代表哈洛德會面，並向他介紹我們的駐點技師伊恩。

 • **I met with Initech's IT manager, Fred Carruthers,** who filled me in on the details concerning **compatibility issues.**
我和創泰公司的資訊科技經理佛萊德・克魯瑟斯會面，他告訴我和相容性問題有關的細節。

## 2 We assessed that the issue was . . .

我們評估問題在於……

- **We assessed that the issue was with a blown fuse.**
  我們評估問題在於燒斷的保險絲。

 • **Preliminary results suggested the problem is the factory's outdated operating system.**
初步結果顯示問題出在該工廠的過時作業系統。

- • **One problem area is the issue of machinery as certain sections need upgrading.**
  出包的範圍是機械的問題，因為某些部分需要升級。

## 3 We offered to . . .

我們提議……

> • We offered to **replace the broken air conditioner and install the new one.**
> 我們提議更換故障的空調並安裝新機。

 相關寫法

• We offered to have discussions with **their staff, ranging from lab technicians to senior management.**
我們提議與他們的員工進行討論，從實驗室技術員到資深管理職都有。

• **Late January** has been provisionally penciled in for follow-up talks.
一月下旬已暫定為後續討論的時間。

---

## 4 After repeated . . ., we found . . .

經過反覆……，我們發現……

> • After repeated **experiments, we found that the display needed to be switched out.**
> 經過反覆實驗，我們發現此顯示器得換掉。

 相關寫法

• Further testing confirmed my suspicions that **the problems were directly related to the motherboard.**
進一步的測試確認了我的疑慮無誤，即問題和主機板有直接的關聯。

157

# 要點式出差報告

**Sass Originals: Sail Center Negotiations Business Trip Report**

Submitted by: Teresa Hogan

Date: June 10

Location: Seoul, South Korea

**Business Trip Events:**

June 4

- Arrived in Seoul at 6:30 p.m. local time
- Checked into Majesty Hotel

June 5

- Met with Lead Buyer Ann Kim at the Sail Center
- Began negotiations to supply them with our designer bags
- Agreed on **wholesale**[1] prices for 500pcs from classic line, 100pcs per style
- **Resumed**[2] discussions after lunch
- Agreed to a 10 percent price **reduction**[3] for an order of 500pcs from the leather **collection**[4]
- Signed a contract for 1,000pcs in total to be delivered by June 28

June 6

- Toured the department store
- **Examined**[5] the location of the future Sass **counter**[6]
- Returned to Taipei, arriving at 9 p.m.

**Follow-up Actions:**

1. Begin recruiting sales crew to **man**[7] the counter
2. Draft and approve layout designs

## 中譯

**薩絲真品：啟航購物中心協商出差報告**

提交人：泰瑞莎・霍根

日期：6 月 10 日

地點：南韓首爾

**出差事項：**

**6 月 4 日**

• 於首爾當地時間晚上 6 點 30 分抵達

• 入住君爵飯店

**6 月 5 日**

• 與採購長金安於啟航購物中心會面

• 開始進行向他們供應設計師包款的協商

• 同意以批發價成交 500 件的經典系列，
每種款式各 100 件

• 午餐過後繼續討論

• 同意 500 件皮革系列的訂單價格打九折

• 簽訂共 1,000 件包款的合約，將於 6 月
28 日前出貨

**6 月 6 日**

• 巡視該百貨公司

• 檢視未來薩絲櫃位的地點

• 返回台北，於晚間 9 點抵達

**後續事項：**

1. 開始招募駐點於櫃台的銷售人員

2. 草擬及批核布置設計

## 字彙

1. **wholesale** [ˋholˏsel] *adj.* 批發的

2. **resume** [rɪˋzum] *v.* （中斷後）繼續

3. **reduction** [rɪˋdʌkʃən] *n.* 減少

4. **collection** [kəˋlɛkʃən] *n.* 系列

5. **examine** [ɪgˋzæmən]
*v.* （仔細地）檢查

6. **counter** [ˋkaʊntə] *n.* 櫃位

7. **man** [mæn] *v.* 為……配備人手

### 延伸學習　出差報告 項目檢測表

• **Trip participant(s)**
出差成員

• **Purpose of travel**
出差目的

• **Summary of completed trip**
行程總結

• **Conclusions & recommendations**
結論和建議

• **Required follow-up actions**
必要的後續行動

• **Additional comments**
附加評論

# 敘述式出差報告

**Quest Corp. Business Trip Report**

To: Leanne Sharp

From: Martin Wells

Date: June 3

## Overview

Destination: Hanoi, Vietnam

Duration of Trip: May 29 to 30

Purpose: Resolve troubles with Delight Textiles' ERP system

## Summary

**Day 1:**

I met with Judy Tran, the HR manager, and introduced her to our senior IT specialist, Mark Johnson. After hearing out Ms. Tran's concerns, we began performing tests on the software. We **assessed**[1] that the issue was with the client's **outdated**[2] operating system and **unfamiliarity**[3] with ERP.

**Day 2:**

We offered to update the client's operating system free of charge. After repeated trials, we found everything to be in working order. Ms. Tran further requested a **remote**[4] training session for the staff there next week as an after-sales service.

## Conclusion

**Faults**[5] with the operating system were **ironed out**.[6] Should any further issues be reported during the training session, I will provide an update.

## 中譯

**探索企業出差報告**
收件人：黎安 · 夏普
寄件人：馬丁 · 威爾斯
日期：6 月 3 日

**概述**
目的地：越南河內
出差期：5 月 29 至 30 日
目的：解決滿欣紡織的企業資源規畫系統問題

**摘要**
第一天：
我與人資經理茱蒂 · 陳會面，並向她介紹我們的資深資料專員馬克 · 強森。聽取陳女士的疑慮後，我們開始進行軟體測試。我們評估問題在於客戶的老舊作業系統以及不熟悉企業資源規畫。

第二天：
我們提議免費升級客戶的作業系統。經過反覆試驗，我們發現一切都正常運作了。陳女士進一步要求下週替當地員工進行遠距訓練課程來做為售後服務。

**結論**
作業系統的故障已排除。訓練課程期間如有通報任何進一步的問題，我將提供更新的狀況。

## 字彙

1. **assess** [əˋsɛs] v. 評估

2. **outdated** [ˌautˋdetəd] adj. 老舊的

3. **unfamiliarity** [ˌʌnfəˌmɪliˋjɛrəti]
   n. 不熟悉

4. **remote** [rɪˋmot] adj. 遠距離的

5. **fault** [fɔlt] n.（機器或系統的）故障

6. **iron out** [ˋaɪən] v. 消除

---

**參考答案**

1. I met with Harry, the sales director, and introduced him to our senior technician, David.

2. We assessed that the issue was with exposure to direct sunlight for extended periods.

3. Preliminary results suggested the problem is the outdated operating system.

4. We offered to conduct further research in order to identify the real problem.

# Thank-You Letters for Business Hospitality
## 出差後謝函

## 課前寫作練習

出差後謝函怎麼寫？
請參考主題詞彙和中譯，將下列句子翻譯成英文。

**❶ 來信旨在誠摯感謝您對我的信心。**

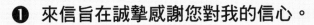

主題詞彙

- **assistance** [əˋsɪstəns]
  協助
- **collaboration**
  [kəˏlæbəˋreʃən] 合作
- **confident** [ˋkɑnfədənt]
  有信心的
- **fruitful** [ˋfrutfəl]
  有成果的

❷ 你真的非常有心,三天晚上都帶我們去高級餐廳吃晚餐。

(寫) _____

_____

❸ 我殷切期盼我們的業務關係能長久且互惠。

(寫) _____

_____

❹ 雖然我有信心全數的相容性問題都已解決,但如果未來您遇到任何軟體失靈的情況,請不吝與我聯絡。

(寫) _____

_____

參考答案請見 p. 169

- **gracious** [ˋgreʃəs]
  殷勤的
- **gratitude** [ˋgrætə͵tud]
  感激之情
- **hospitality** [͵hɑspəˋtæləti]
  款待
- **reception** [rɪˋsɛpʃən]
  接待

## 1 I'm writing to express my deepest gratitude for . . .

來信旨在誠摯感謝……

- **I'm writing to express my deepest gratitude for your gracious hospitality during my brief visit.**
  來信旨在誠摯感謝您在我短暫拜訪期間的殷勤款待。

 相關寫法
- I would like to say how much I enjoyed meeting you in Chicago last week.
  我想表達我非常開心上週在芝加哥與您會面。

---

## 2 I appreciate the time you took to . . .

感謝您抽空……

- **I appreciate the time you took to show me around the city.**
  感謝您抽空帶我遊覽整座城市。

 相關寫法
- It was very kind of you to **take our team out for a wonderful meal.**
  您真的非常有心，帶我們的小組成員去外面吃了頓大餐。

- You were especially detailed in **your explanations of Bellwood Corporation's position.**
  您對貝伍德公司情況的說明尤其詳盡。

## 3　I have high hopes that . . .

我殷切期盼⋯⋯

> • **I have high hopes that** our cooperation will lead to significant returns for your company.
> 我殷切期盼我們的合作將為貴司帶來可觀的報酬。

 • **I'm extremely hopeful that** our business relationship will be fruitful.
我殷切期盼我們的業務關係會有豐碩的成果。

• **I believe that** we will be able to work out a satisfactory partnership agreement.
我相信我們將能討論出令人滿意的合作協議。

---

## 4　Although I'm confident that . . ., please don't hesitate to . . .

雖然我有信心⋯⋯，請不吝⋯⋯

> • **Although I'm confident that** you'll be able to handle any further issues, **please don't hesitate to** reach out to me if you need assistance.
> 雖然我有信心您將能處理任何進一步的問題，但如果您需要協助，請不吝與我聯繫。

 • **If,** at any time, I might be of further assistance, please call upon me.
不論何時，只要我能提供進一步的協助，請與我聯繫。

# 洽談合作篇

To: rachel.kirkpatrick@thefoundry.com
From: drew.buchanan@domaindesigns.com
Subject: Look Forward to Working Together

Dear Rachel,

I would like to say how much I enjoyed meeting you in San Francisco. I believe that we were able to **flesh out**[1] a clear and **practical**[2] picture of the layout for your future office. Fortunately, your great taste in decorations made the process a **breeze**.[3]

I'd also like to thank you for your hospitality. It was very kind of you to show me around the city. Hopefully, I will get the chance to do the same the next time you are in Los Angeles.

> 形容「相信未來會發生好事的希望」，亦即「殷切期盼」。

I have high hopes that our first meeting will **blossom**[4] into a fruitful collaboration. I look forward to our continuing cooperation.

Sincerely,

Drew Buchanan
Interior Designer
Domain Designs Ltd.

## 中譯

瑞秋你好，

我想表達我非常開心在舊金山與你會面。我相信我們能構築出你未來辦公室清楚且實際的格局樣貌。很幸運地，你對裝潢的絕佳品味讓討論的過程相當容易。

我也要謝謝你的款待。你真的非常有心，帶我遊覽整座城市。但願下次你來洛杉磯時，我也有機會以禮相待。

我殷切期盼我們的首次會晤能結出豐碩的合作成果。期待我們的持續配合。

謹啟，

德魯・布坎南
室內設計師
域界設計有限公司

## 字彙

1. **flesh out** [flɛʃ]
   v. （以細節或資訊）充實

2. **practical** [ˋpræktɪkəl] adj. 實際的

3. **breeze** [briz] n. 輕而易舉之事

4. **blossom** [ˋblɑsəm] v. 深入發展

---

### 延伸學習　出差後謝函 寫作要點

- **As you express thanks, mention the event or circumstance involved.**
  表達謝意時，提及相關的事件或情況。

- **Add specific comments about what you enjoyed.**
  對於所喜愛的事項補充特別的評論。

- **Mention details about the reader's thoughtfulness.**
  詳述收信者的貼心之處。

- **Offer to reciprocate in the future.**
  期望未來也能以禮相待。

# 問題排除篇

To: allison.osborne@takefive.com
From: william.potter@shiftsmartoffice.com
Subject: Subject Kind Thanks for Your Help

Dear Ms. Osborne,

I'm writing to express my deepest gratitude for the help you provided while I was at your office. I appreciate the time you took to clearly describe the issues with our hardware.

Without your timely assistance, I would not have been able to **define**[1] and solve the problems so quickly. I would just remind you that any cables or power **circuits**[2] attached to the smart-office **setup**[3] should not be touched.

Although I'm confident that the issues have been **resolved**,[4] please don't hesitate to call me should anything **arise**.[5] I'd be delighted to provide any follow-up services.

Regards,

William Potter
On-site Technician
Shift Smart Office Inc.

## 中譯

奧斯柏恩小姐您好，

來信旨在誠摯感謝您在我造訪貴司時所提供的幫助。感謝您抽空清楚描述敝司硬體的問題。

沒有您的即時協助，我就無法如此快速地界定和解決問題了。我只是要提醒您，固定在智慧辦公室裝置的任何電纜或電路皆不可碰觸。

雖然我有信心這些問題已解決了，但若有任何狀況發生，請不吝打電話給我。我很樂意提供任何後續的服務。

謹致，

威廉·波特
駐點工程師
席福特智慧辦公公司

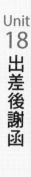

## 字彙

1. **define** [dɪˋfaɪn] v. 界定；確定

2. **circuit** [ˋsɝkət] n. （電流的）電路

3. **setup** [ˋsɛt͵ʌp] n. 裝置

4. **resolve** [rɪˋzɑlvt] v. 解決

5. **arise** [əˋraɪz] v. 出現

**參考答案**

1. I'm writing to express my deepest gratitude for your confidence in me.

2. It was very kind of you to take us to dinner all three evenings at the most elegant restaurants.

3. I have high hopes that our business relationship will be a long and mutually beneficial one.

4. Although I'm confident that all of the compatibility issues have been resolved, please don't hesitate to contact me if you encounter any future glitches.

# Sales Reports
## 業績報告

## 課前寫作練習

業績報告怎麼寫？
請參考主題詞彙和中譯，將下列句子翻譯成英文。

**❶** 公司海外營運處的銷售淨額增加 42%，從 2019 年的 290 萬美元上升至 2020 年的 410 萬美元。

 _____

_____

主題
詞彙

- **decline/dip** [dɪˋklaɪn] [dɪp]
  下滑；減少
- **highlight** [ˋhaɪˌlaɪt]
  摘要

- **net sales** [nɛt]
  銷售淨額
- **pick up**
  提升

❷ 第三季的銷量下滑，絕大部分是因為夏末的庫存短
缺。

(寫) _____

_____

❸ 公司的防禦性行銷活動結果相當成功，在第四季的
銷售增幅當中占了將近 25%。

(寫) _____

_____

❹ 我們預估新的子公司在 2022 年將會有 10% 的年
成長率。

(寫) _____

_____

參考答案請見 p. 177

- **profit/return** [ˈprɑfət] [rɪˈtɜn]
  獲益
- **project** [prəˈdʒɛkt]
  預估

- **rebound** [ˈriˌbaʊnd]
  回升；反彈
- **revenue** [ˈrɛvəˌnu]
  營收

## 1 (Company) saw net sales increase by (percent) from (amount) in (year) to (amount) in (year).

（公司）的銷售淨額增加（百分比），從（年分）的（金額）上升至（年分）的（金額）。

- **Plot Ventures** saw net sales increase by **23 percent** from **$18 million** in **2019** to **$22.1 million** in **2020.**

  普拉特創投的銷售淨額增加 23%，從 2019 年的 1,800 萬美元上升至 2020 年的 2,210 萬美元。

 相關寫法
- In its **third year**, the online division of Globex grew by **20 percent** in sales from **$15 million** to **$18 million.**

  格羅拜斯公司的網路事業處，第三年的銷售額成長 20%，從 1,500 萬美元上升至 1,800 萬美元。

---

## 2 (Quarter) was customarily slow with sales picking up in (quarter).

（季度）成長力道一如往常地緩慢，銷量至（季度）才攀升。

- **Q1** was customarily slow with sales picking up in **Q2.**

  第一季成長力道一如往常地緩慢，銷量至第二季才攀升。

 相關寫法
- Returns in **Q4** dipped slightly as **demand in China fell.**

  由於中國的需求降低，第四季的利潤微幅下滑。

- **Q3** sales rebounded, largely due to **the company's revamped marketing strategy.**

  第三季的銷量回升，絕大部分是因為公司調整後的行銷策略。

## 3 (Campaign) proved a resounding success, accounting for (percent) of the increase in sales in (quarter).

（活動）結果相當成功，在（季度）的銷售增幅當中占（百分比）。

- **The "Buy 1, Gift 1" book donation campaign** proved a resounding success, accounting for **15 percent** of the increase in sales in **Q3.**

  「買一送一」的捐書活動結果相當成功，在第三季的銷售增幅當中占 15%。

 • **Q2** sales bounced back mostly because of **the company's new ad campaign.**

第二季的銷量得以回升，主因在於公司的新廣告活動。

---

## 4 We project that (company) will be able to increase its annual growth rate by (percent) in (year).

我們預估（公司）在（年分）將會有（百分比）的年成長率。

- We project that **Kon Capital** will be able to increase its annual growth rate by **12 percent** in **2021.**

  我們預估康恩資本公司在 2021 年將會有 12% 的年成長率。

 • **Delos Inc.** is projected to double its growth to **14 percent** in **2021.**

戴洛斯公司的成長幅度預計在 2021 年增加一倍達 14%。

- **We expect that Tyrell Corp.** will be able to boost its annual growth in **2021** to **16 percent.**

我們預期泰洛公司 2021 年的年成長幅度能提高到 16%。

Unit
19
業績報告

# 服飾品牌業績報告

## Blurred Apparel 2020 Annual Sales Report

### Clifford Lee, CFO, Blurred Apparel

> 為 **chief financial officer**「財務長」首字縮略語。

Blurred Apparel saw net sales increase by 12 percent from $91.3 million in 2019 to $103.8 million in 2020. During this period, Blurred Apparel profited from its high-quality and low-cost **offerings**[1] and well-executed **multichannel**[2] sales strategy.

### Financial highlights of 2020

Q1 was **customarily**[3] slow with sales picking up in Q2. Unfortunately, returns in Q3 dipped slightly as U.S. **tariffs**[4] began to **impact**[5] clothing prices. Q4 sales rebounded, largely due to the company's **aggressive**[6] holiday promotions. The online year-end sale single-handedly accounted for 8 percent of increased revenues.

Blurred Apparel's expanding online presence and men's clothing line present significant growth opportunities for the company. Based on the brand's current **momentum**,[7] we project that Blurred Apparel will be able to increase its annual growth rate by 30 percent in 2021.

### 布勒德服飾 2020 年年度業績報告
克里弗德‧李，布勒德服飾財務長

布勒德服飾的銷售淨額增加 **12%**，從 **2019** 年的 **9,130** 萬美元上升至 **2020** 年的 **1.038** 億美元。在此期間，布勒德服飾獲利於其高品質且低成本的出售品項，以及執行得當的多通路銷售策略。

#### 2020 年財務摘要

第一季成長力道一如往常地緩慢，銷量至第二季才攀升。可惜的是，由於美國關稅開始衝擊服飾價格，第三季的利潤微幅下滑。第四季的銷量回升，絕大部分是因為公司積極的假期促銷活動。歲末網路銷售獨占增加營收的 **8%**。

布勒德服飾拓展中的網路能見度與男性服飾產線，展現了公司的重要成長契機。基於品牌目前的動力，我們預估布勒德服飾在 2021 年將會有 **30%** 的年成長率。

## 字彙

1. **offering** [ˋɔfərɪŋ] *n.* 出售品項

2. **multichannel** [͵mʌltiˋtʃænl̩] *adj.* 多通路的

3. **customarily** [͵kʌstəˋmɛrəli] *adv.* 如往常地

4. **tariff** [ˋtɛrəf] *n.* 關稅

5. **impact** [ɪmˋpækt] *v.* 衝擊；影響

6. **aggressive** [əˋgrɛsɪv] *adj.* 雄心勃勃的

7. **momentum** [moˋmɛntəm] *n.* 動力

---

### 延伸學習　業績報告 組成元素

- **report title**
  報告標題

- **author's name**
  作者姓名

- **overall sales performance**
  整體業績表現

- **respective quarterly results**
  各季度成果

- **future goals**
  未來目標

# 網路商城業績報告

### Pages Online 2020 Annual Sales Report

Buster Olson, CFO, Pages Online

In its second year, the online division of Pages grew by 23 percent in sales from US$15.2 million to US$18.7 million, **surpassing**[1] its goal by 4.2 percent. As sales of traditional books are declining, Pages Online continues to successfully expand into the e-publishing market as well as a growing **array**[2] of lifestyle products.

### Financial highlights 2020

Sales picked up in Q2 after a typically slow Q1, but the economic **downturn**[3] in mid-2020 caused a sharp decline in Q3 figures. To **remedy**[4] the situation, several key promotions were **implemented**.[5] The "Buy 1, Gift 1" book donation campaign proved a **resounding**[6] success, accounting for 45 percent of the increase in sales in Q4, re-stabilizing profits to projected levels.

> 字根 -conscious 指「看重……的」，
> quality-conscious 即指「注重品質的」。

Pages Online is collaborating with local designers to broaden the Pages-At-Home line of above-average lifestyle products and gifts, targeting quality-conscious consumers. This department is projected to double its growth to 12 percent in 2021, while the overall projected growth is set at a relatively ambitious 30 percent.

## 佩吉斯線上公司 2020 年年度業績報告
### 巴斯特・歐森，佩吉斯線上公司財務長

佩吉斯公司的網路事業處，第二年的銷售額成長 **23%**，從 **1,520** 萬美元上升至 **1,870** 萬美元，超越目標達 **4.2%**。隨著傳統書籍的銷量下滑，佩吉斯線上公司持續成功地拓展至電子出版市場，同時推出越來越多的生活用品。

### 2020 年財務摘要

經過第一季經常性的表現遲緩，銷售情況在第二季出現起色，但 **2020** 年年中的經濟衰退導致第三季的銷售數字大幅下滑。為補救此一情形，公司執行了幾個關鍵的促銷活動。「買一送一」的捐書活動結果相當成功，在第四季的銷售增幅當中占 **45%**，使利潤重新回穩至預估水平。

佩吉斯線上公司正與本土設計師合作，擴展佩吉斯居家系列的高檔生活用品和禮品，以注重品質的消費者為目標。此部門的成長幅度預計在 **2021** 年增加一倍達 **12%**，而整體的預估成長則設定在相當有野心的 **30%**。

## 字彙

1. **surpass** [səˋpæs] *v.* 超越

2. **array** [əˋre] *n.* 大量

3. **downturn** [ˋdaʊn͵tɝn] *n.* 經濟衰退

4. **remedy** [ˋrɛmədi] *v.* 補救

5. **implement** [ˋɪmplə͵mɛnt] *v.* 實施

6. **resounding** [rɪˋzaʊndɪŋ] *adj.* 驚人的

---

參考答案

1. Our overseas operations saw net sales increase by 42 percent from $2.9 million in 2019 to $4.1 million in 2020.

2. Q3 sales declined, largely due to the shortage of inventory in late summer.

3. The company's defensive marketing campaign proved a resounding success, accounting for nearly 25 percent of the increase in sales in Q4.

4. We project that the new subsidiary will be able to increase its annual growth rate by 10 percent in 2022.

# Social Interactions
## 社交公關篇

# Unit 20 Business Invitations
## 商務邀請函

## 課前寫作練習

商務邀請函怎麼寫？
請參考主題詞彙、中譯及括號內的英文提示，將下列句子翻譯成英文。

❶ 謹代表義大利研究協會（Italian Studies Society），我想邀請您蒞臨文藝復興藝術研討會（Renaissance Art Conference），預訂從 3 月 13 日至 15 日假希爾頓飯店（Hilton Hotel）舉行。

_____

_____

- **attendance** [əˋtɛndəns]
  到場；出席
- **attendee** [ə.tɛnˋdi]
  出席者

- **honor** [ˋɑnɚ]
  榮幸
- **invitee** [.ɪnvaɪˋti]
  受邀者

❷ 誠摯邀請您參加即將於 6 月 10 日至 12 日假 CBT 會議中心（CBT Conference Center）所舉辦的台菲商務論壇（Taiwan-African Business Forum）。

（寫）

_____

_____

❸ 我們深感榮幸您能擔任我們的主講者。

（寫）

_____

_____

❹ 煩請於 10 月 15 日前回覆此邀請。

（寫）

_____

_____

參考答案請見 p. 187

- **on behalf of** [bɪˋhæf]
  謹代表
- **participation** [pɑrˌtɪsəˋpeʃən]
  參與

- **RSVP** 敬請回覆（源自法語的同義說法 répondez s'il vous plaît）
- **schedule** [ˋskɛdʒul]
  安排；排定

TRACK 060

## 1 On behalf of (organization), I would like to invite you to (occasion) scheduled from (date) to (date) at (location).

謹代表（組織），我想邀請您蒞臨（盛事），預訂從（日期）至（日期）假（地點）舉行。

- On behalf of **the Taiwan Cycling Association,** I would like to invite you to **the Taipei Bike Expo** scheduled from **August 5** to **August 7** at **the First Convention Center.**

  謹代表台灣自行車協會，我想邀請您蒞臨台北自行車展，預訂從 8 月 5 日至 7 日假第一會議中心舉行。

 • You are cordially invited to **the Solar Tech Conference** that will be held from **July 5** to **7** at **the Ritz-Carlton.**

  誠摯邀請您蒞臨即將於 7 月 5 日至 7 日假麗思卡爾頓酒店所舉辦的太陽能科技研討會。

## 2 It would be our honor to . . .

我們深感榮幸……

- It would be our honor to **have you lead the opening ceremony.**

  我們深感榮幸您能來主領開幕典禮。

 • We would love it if you, **a valued member of our community,** could grace us with your presence.

  若您這位團體的重要成員能蒞臨為我們增光，我們將非常開心。

## 3 We will be offering . . .

我們將提供……

- We will be offering **invitees a gift bag filled with our top-of-the-line products.**

  我們將提供受邀者一個裝滿敝司頂級產品的福袋。

 • The conference arranged by **the Solar Power Alliance** provides a forum for discussion on **hot industry trends.**

  這個由太陽能聯盟所籌備的研討會提供一個論壇讓大家討論熱門產業趨勢。

## 4 Please kindly respond to this invitation by (date).

煩請於（日期）前回覆此邀請。

- Please kindly respond to this invitation by **August 26.**

  煩請於 8 月 26 日前回覆此邀請。

 • Please RSVP by **November 1** to confirm your attendance at **the conference.**

  請在 年 11 月 1 日以前回覆以確定您將出席此研討會。

- Please let me know by **August 21** if you are available that day so we may **reserve your spot.**

  請於 8 月 21 日前讓我知道您當天是否有空，以便我們預留您的席位。

# 商務會議邀請函

To: morgan.cobb@skylink.com
From: christian.hudson@allthingsai.com
Subject: Global AI Conference Keynote

Dear Ms. Cobb,

On behalf of All Things AI, I would like to invite you to the 5th annual Global AI Conference scheduled from September 23 to 27 at the San Francisco Regatta Hotel.

Many leaders from the tech industry will be in attendance to share their **perspectives**[1] on recent developments in AI. You can expect the likes of the CEOs of Blitz and RainCloud to speak on our **panels**.[2]

> the likes of sb/sth 指「類似……的人事物」。

It would be our honor to have your participation in the conference as the **keynote speaker**.[3] We believe your **invaluable**[4] **insights**[5] on the future of AI would be of great benefit to our attendees.

We would be delighted if you could join us. Please kindly respond to this invitation by August 12.

Best regards,

Christian Hudson
Conference Coordinator
All Things AI

## 中譯

卡柏小姐您好，

謹代表萬事人工智慧，我想邀請您蒞臨第五屆年度全球人工智慧大會，預訂從 9 月 23 日至 27 日假舊金山里加塔飯店舉行。

眾多來自科技界的領袖都將到場分享他們對人工智慧近期發展的觀點。您可以預期閃電公司和雨雲公司的執行長等人物在我們的專題討論小組發表談話。

我們將深感榮幸您能以主講者的身分參與此研討會。我們相信您對人工智慧未來的寶貴洞見能為與會者帶來諸多裨益。

若您能一同加入，我們會非常開心的。煩請於 8 月 12 日前回覆此邀請。

謹致，

克里斯汀・哈德森
會議籌劃專員
萬事人工智慧

## 字彙

1. **perspective** [pɚ`spɛktɪv] *n.* 觀點

2. **panel** [`pænl̩] *n.* 專題討論小組

3. **keynote speaker** [`ki͵not]
   *n.* 主講者

4. **invaluable** [ɪn`væljəbəl]
   *adj.* 寶貴的

5. **insight** [`ɪn͵saɪt] *n.* 洞見

---

**延伸學習** **商務邀請函 組成元素**

- **Special offers or incentives**
  特別的提議或動機

- **Event schedule and location**
  活動時程和地點

- **Catering information**
  餐飲資訊

- **RSVP information**
  回覆資訊

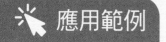

# 商務活動邀請函

To: neal.brewer@chocolatier.com
From: sandra.mccarthy@theparisian.com
Subject: Grand Opening!

Dear Mr. Brewer,

You are **cordially**[1] invited to the grand opening of our third location in Taipei. We would love it if you, one of our most **esteemed**[2] suppliers, could join us for the celebrations.

We will be offering all of our guests a free box of our most delicious **pastries**.[3] There will also be food, wine, and live music. The event will run throughout the whole day.

The time, date, and location are as follows:

• Start Time: 11 a.m.

• Date: August 30

• Location: No. 186, Lane 19, Guangfu South Road, Xinyi District, Taipei City

This opening **marks**[4] a **momentous**[5] **milestone**[6] in our short history. We are excited to share in the **festivities**[7] with you. Please let me know by August 9 if you are available that day so we may prepare your gift box.

Yours sincerely,

Sandra McCarthy
Branch Manager
The Parisian Patisserie

## 中譯

布魯爾先生您好，

誠摯邀請您參加我們台北第三個據點的盛大開幕。若您這位我們至為敬重的供應商能一同參與此慶祝活動，我們將非常開心。

我們將提供所有賓客一盒極為可口的免費自家茶點。也會有食物、酒水及現場演奏。活動將進行一整天。

時間、日期與地點如下：
- 開始時間：上午十一點
- 日期：8 月 30 日
- 地點：台北市信義區光復南路 19 巷 186 號

此開幕式在我們短短的店史中標誌為重要的里程碑。我們盛情與您分享這些慶祝活動。請於 8 月 9 日前讓我知道您當天是否有空，以便我們準備您的禮盒。

謹啟，

珊卓・麥卡錫
分店經理
巴黎人法式甜點

## 字彙

1. **cordially** [ˋkɔrdʒəlɪ] *adv.* 誠摯地

2. **esteemed** [ɪˋstimd] *adj.* 敬重的

3. **pastry** [ˋpestrɪ] *n.* 茶點

4. **mark** [mɑrk] *v.* 標誌

5. **momentous** [moˋmɛntəs] *adj.* 重要的

6. **milestone** [ˋmaɪl͵ston] *n.* 里程碑

7. **festivity** [fɛˋstɪvətɪ] *n.* 慶祝活動

**參考答案**

1. On behalf of the Italian Studies Society, I would like to invite you to the Renaissance Art Conference scheduled from March 13 to 15 at the Hilton Hotel.

2. You are cordially invited to the Taiwan-African Business Forum that will be held from June 10 to 12 at the CBT Conference Center.

3. It would be our honor to have you as our keynote speaker.

4. Please kindly respond to this invitation by October 15.

# Letters of Appreciation
# 感謝信

## 課前寫作練習

感謝信怎麼寫？
請參考主題詞彙和中譯，將下列句子翻譯成英文。

❶ 謹代表敝司特此致函，向您獻上最深的謝意，感謝您能體諒最近我們在運送方面的延誤。

（寫）

_____

_____

主題詞彙

- **assistance** [əˋsɪstəns]
  協助
- **contribution**
  [ˌkɑntrəˋbjuʃən] 貢獻

- **grateful** [ˋgretfəl]
  感激的
- **gratitude** [ˋgrætəˌtud]
  感謝

❷ 來信是為了感謝有這個機會能提供我們的產品給貴司。

⟨寫⟩

_____

_____

❸ 當我正苦於籌畫會議時,是你為我提供了清晰的見解和組織技巧。

⟨寫⟩

_____

_____

❹ 為表達謝意,我們想在下週請你去吃午餐。

⟨寫⟩

_____

_____

參考答案請見 p. 195

- **kindness** [ˈkaɪndnəs]
  好意;體貼
- **thoughtful** [ˈθɔtfəl]
  關心的

- **token** [ˈtokən]
  表示;象徵
- **value** [ˈvælju]
  重視

## 1 I am writing on behalf of my company to express our deepest gratitude for . . .

謹代表敝司特此致函，向您獻上最深的謝意，感謝您……

- I am writing on behalf of my company to express our deepest gratitude for **being chosen as the sole beverage distributor to restaurants part of the Samatha's Chicken franchise.**

  謹代表敝司特此致函，向您獻上最深的謝意，感謝您選擇我們做為莎曼珊炸雞連鎖餐廳的單一飲料經銷商。

 • I am writing to express my appreciation for **all of the helpful advice you've given me over the past month.**

  來信是為了感謝您在過去這個月給予我的所有實用建議。

## 2 I am so grateful for . . .

我非常感激……

- I am so grateful for **your guidance during the busiest sales quarter to date.**

  我非常感激您從銷售旺季一直到今天所給予我的指導。

 • I'd like to offer you my sincere thanks for **stopping by our booth at the recent trade show.**

  您在最近的商展上蒞臨我們的攤位，對此我獻上我誠摯的謝意。

## 3 While I was struggling to . . ., you . . .

當我正苦於……時，是你……

- While I was struggling to **understand the new data entry system,** you **stayed after work to teach me how to use it.**

  當我正苦於瞭解新的資料輸入系統時，是你下班後留下來教我如何使用的。

 • I could not have **met this month's deadline** without your **market insights and sales tips.**

  沒有你的市場洞見和銷售訣竅，我就無法趕上這個月的期限了。

- You cannot imagine what a help this was to me as **I would not otherwise have been able to meet my deadline.**

  你無法想像這個援助對我的意義有多大，不然我很有可能會趕不上截止期限。

## 4 As a token of our appreciation, we . . .

為表達謝意，我們……

- As a token of our appreciation, we **would like to offer Dakang Industries a 20 percent discount on the next order.**

  為表達謝意，我們想提供大康工業下筆訂單 20% 的折扣。

 • To say thank you in a more concrete way, we would like to **offer you a special gift.**

  為了更具體展現我們的謝意，我們想提供您一份特別的禮物。

# 感謝客戶篇

To: asanderson@lamdengolf.com
From: kwagstaff@kesonsportinggoods.com
Subject: Gratitude for Shipping Extension

Dear Mr. Sanderson,

My name is Ken Wagstaff. I am the senior executive of Sales, writing on behalf of my company to express our deepest gratitude for **extending**[1] the shipping deadline on your order of Titanium R18 596 Drivers. The **delay**[2] in production at our manufacturing plant was caused by a technical **malfunction**,[3] which has since been corrected.

We value the long term business that Lamden Golf Equipment has provided and are **committed**[4] to **delivering**[5] quality products at the most competitive factory prices. As a token of our appreciation for your patience, we are pleased to offer you a 10 percent discount on your next order. We look forward to working with you again in the future.

Sincerely,

Ken Wagstaff
Senior Sales Executive
Keson Sporting Goods Co.

## 中譯

山德森先生您好，

我的名字是肯恩·瓦格史戴夫。我是業務部的資深主管，謹代表敝司特此致函，向您獻上最深的謝意，感謝您願意延後 **R18 596** 鈦質球桿的出貨期限。工廠的生產延遲是由技術性的故障所造成，如今業已修正。

我們相當重視蘭登高爾夫球設備公司長期提供的業務，並致力以最具競爭力的工廠價格來做出高品質的產品。為表達對您耐心等候的謝意，我們非常樂意提供您下筆訂單 **10%** 的折扣。期待未來再次與您合作。

謹啟，

肯恩·瓦格史戴夫
資深業務經理
凱森運動用品公司

## 字彙

1. **extend** [ɪk`stɛnd] *v.* 延長

2. **delay** [dɪ`le] *n.* 延遲

3. **malfunction** [ˌmæl`fʌŋkʃən] *n.* 故障

4. **committed** [kə`mɪtəd] *adj.* 致力的

5. **deliver** [dɪ`lɪvə] *v.* 實現；產生

 延伸學習 **感謝信 寫作要點**

- State what you appreciate.
  說明所感謝的內容。

- Be specific about the person's work or actions.
  具體說明對方的作為或行動。

- Be brief, warm, and sincere.
  要簡短、熱情且真誠。

- Be slightly formal.
  要正式一點。

# 感謝同事篇

---

To: mburns@kesonsportinggoods.com

From: kwagstaff@kesonsportinggoods.com

Subject: Appreciation for Your Assistance

---

Dear Marcia,

I am writing to express my appreciation for your assistance in **gathering**[1] technical and market-based information on the new Titanium R18 596 Drivers. I am so grateful for all the hard work you have done. While I was **struggling**[2] to find **adequate**[3] data in relation to manufacturing costs and consumer demand for my sales presentation, you helped a great deal even though you had a full workload yourself.

> 指「和⋯⋯有關」，意同 in connection to。

Because of your efforts, the product **pitch**[4] was very **well received**[5] by many key distributors, who then placed large orders. In order to return your kindness, I would like to treat you to dinner this Friday if you are free. I could not have made such an **informed**[6] presentation without your contributions.

Thank you very much,

Ken

## 中譯

瑪夏你好，

來信是為了感謝你協助蒐集新 **R18 596** 鈦質球桿的技術面與市場面資料。我非常感激你所有的付出。當我為了業務簡報而苦於找齊和製造成本及消費者需求相關的資料時，儘管你自己工作滿載，卻還是傾力相助。

由於你的付出，產品簡報廣受許多重要經銷商的好評，他們也因此下了大筆訂單。為了報答你的好意，若你這週五有空，我想請你吃晚餐。沒有你的貢獻，我就無法做出內容如此豐富的簡報了。

非常謝謝你，

肯恩

## 字彙

1. **gather** [ˋgæðɚ] *v.* 蒐集

2. **struggle** [ˋstrʌgəl] *v.* 努力

3. **adequate** [ˋædɪkwət]
   *adj.* 足夠的

4. **pitch** [pɪtʃ] *n.* 簡報；投售

5. **well received** [rɪˋsivd] 深受好評的

6. **informed** [ɪnˋfɔrmd]
   *adj.* 資訊充足的

**參考答案**

1. I am writing on behalf of my company to express our deepest gratitude for your understanding about the recent shipping delays.

2. I am writing to express my appreciation for having the opportunity to supply your company with our products.

3. While I was struggling to organize the meeting, you provided me with your lucid insights and organizational skills.

4. As a token of our appreciation, we would like to invite you out to lunch next week.

# Congratulation Letters
## 祝賀信

## 課前寫作練習

祝賀信怎麼寫？
請參考主題詞彙、中譯及括號內的英文提示，將下列句子翻譯成英文。

❶ 特此致函恭賀你的新事業，安德魯斯顧問公司
（Andrews Consulting）的盛大開幕。

---

---

主題詞彙

- **achievement** [əˋtʃivmənt]
  成就；成績
- **admire** [ədˋmaɪr]
  欣賞

- **congratulate**
  [kənˋgrætʃəˌlet] 祝賀
- **contribution**
  [ˌkɑntrəˋbjuʃən] 貢獻

❷ 請接受我對你成就斐然最衷心的祝賀。

(寫) _____

_____

❸ 我很高興得知你被委派了這項重要的任務。

(寫) _____

_____

❹ 你的足智多謀（ingenuity）與堅持不懈（perseverance）總是讓我印象深刻。

(寫) _____

_____

參考答案請見 p. 203

- **deserve** [dɪˋzɝv]
  值得；應得
- **fantastic** [fænˋtæstɪk]
  極好的

- **impress** [ɪmˋprɛs]
  使欽佩
- **promotion** [prəˋmoʃən]
  升遷

## 1 I am writing to congratulate you on . . .

特此致函恭賀你⋯⋯

- **I am writing to congratulate you on your advancement to regional manager.**
  特此致函恭賀你晉升為區域經理。

 相關寫法
- I take great pleasure in sending congratulations to you about **your promotion to division manager.**
  我很開心要祝賀你晉升為事業處經理。

- Please accept my heartiest congratulations on **receiving the Outstanding Service Award.**
  請接受我對你獲得傑出服務獎最衷心的祝賀。

## 2 I am writing to let you know how much your work on . . . is appreciated.

特此致函讓你知道，我很感激你在⋯⋯上所做的努力。

- **I am writing to let you know how much your work on the Blake project is appreciated.**
  特此致函讓你知道，我很感激你在布雷克專案上所做的努力。

 相關寫法
- I am in awe of all that you've done on **the Anderson project.**
  我對你在安德森專案上所做的一切感到驚嘆。

## 3　I am delighted to hear that you . . .

我很高興得知你……

> 為 chief operating officer「營運長」的首字縮略語。

- I am delighted to hear that you **have been named COO at Layton Associates.**

  我很高興得知你被任命為雷頓聯合公司的營運長。

 • I wanted you to know how proud and happy I was to hear that **you have struck a deal with the big company.**

我想讓你知道在我聽到你與這家大公司敲定合約時，我感到多麼驕傲和開心。

---

## 4　I've always been impressed with your . . .

你的……總是讓我印象深刻。

- I've always been impressed with **your patience and carefulness.**

  你的耐心與細心總是讓我印象深刻。

 • You have done a fantastic job **completing the joint venture proposal.**

你漂亮地完成了這項合資事業提案。

• I very much admire your **organizational skills and the high quality of your work.**

我非常欽佩你的組織能力和高工作品質。

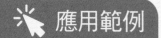 

# 祝賀升遷

To: jwhite@aghatch.com
From: gpearce@austinsolution.com
Subject: Congratulations on Your Promotion

Dear Julie,

I am writing to congratulate you on your recent promotion. I am delighted to hear that you have been named the new sales manager at AG Hatch.

> come as no surprise 指「毫不意外」。

This announcement certainly comes as no surprise to me. I've always been impressed with your hard work and **dedication**[1] to your company. You definitely deserve this new title, and I am sure that you will **excel**[2] in this new position.

Once again, congratulations. I wish you continued success in your career.

Sincerely,

Guy Pearce
President
Austin Solutions

## 中譯

親愛的茱莉，

特此致函恭賀你最近獲得升遷。我很高興得知你被任命為 **AG** 企畫公司的新業務經理。

這項通知對我來說當然毫不意外。你的努力和對公司的貢獻總是讓我印象深刻。你絕對值得這個新頭銜，而我也確信你在這個新職位會有優異的表現。

再次恭喜你。祝你職涯一帆風順。

謹啟，

蓋‧皮爾斯
總裁
奧斯汀解決方案公司

## 字彙

*1.* **dedication** [ˌdɛdəˈkeʃən] *n.* 貢獻

*2.* **excel** [ɪkˈsɛl] *v.* （表現）突出

## 延伸學習　祝賀信 寫作要點

- **Mention the reasons to congratulate.**
  提及祝賀的理由。

- **Tell how you learned about the good news.**
  告知你如何得知好消息。

- **Describe how happy you are and why.**
  描述你有多開心和原因。

- **Wish the person continued success and happiness.**
  祝福對方一帆風順、幸福快樂。

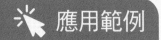 

# 恭賀出色表現

To: jdowns@kentfinancialservices.com
From: hminton@gspmanufacturing.com
Subject: Congratulations on a Job Well Done

Dear Jane,

I am writing to let you know how much your work on the year-end budget report is appreciated. You have done a fantastic job completing this difficult and **time-consuming**[1] assignment under a tight deadline. I know how much effort goes into completing such a report, and your dedication is **apparent**[2] in every page of this document.

> **go into** 指「（時間、金錢、精力等）花在……上、投入……」。

Others in my company have also had high **praise**[3] for your work. We are glad that we have had the chance to work with you on this, and we look forward to continued cooperation in the future.

Sincerely,

Hal Minton
Managing Director
GSP Manufacturing

## 中譯

親愛的珍，

特此致函讓你知道，我很感激你在年終預算報告上所做的努力。你在緊迫的期限內漂亮地完成了這項艱鉅且耗時的任務。我明白完成這樣的報告需要投入多少的努力，而你的貢獻在這份文件上的每一頁都不言而喻。

我公司的其他同仁也都對你的努力高度讚賞。我們很高興能有這個機會和你一起合作此案，也期盼未來能繼續合作。

謹啟，

哈爾‧明頓
常務董事
**GSP** 製造公司

## 字彙

*1.* **time-consuming**
[ˈtaɪmkənˌsumɪŋ] *adj.* 耗時的

*2.* **apparent** [əˈpɛrənt]
*adj.* 顯而易見的

*3.* **praise** [prez] *n.* 讚揚

參考答案

1. I am writing to congratulate you on the grand opening of your new business, Andrews Consulting.

2. Please accept my heartiest congratulations on your remarkable achievement.

3. I am delighted to hear that you have been delegated this important task.

4. I've always been impressed with your ingenuity and perseverance.

# Acknowledgments and Recognitions
## 表揚信

## 課前寫作練習

表揚信怎麼寫？
請參考主題詞彙、中譯及括號內的英文提示，將下列句子翻譯成英文。

❶ 我想藉此機會，對你的努力表達誠摯的謝意。

_____

_____

主題詞彙

- **capability**
  [ˌkepəˋbɪlətɪ] 產能

- **competence**
  [ˋkɑmpətəns] 能力

- **coordination**
  [koˌɔrdəˋneʃən] 協調

- **dependability**
  [dɪˌpɛndəˋbɪlətɪ] 可靠

❷ 值得注意的是你為了防止可能的數據損失所做的一切。

(寫)

_____

_____

❸ 我很感謝你願意為同事盡心盡力（go the extra mile）。

(寫)

_____

_____

❹ 為表達我們對你所貢獻的感激之情，我們想給你一個月薪資的獎金。

(寫)

_____

_____

參考答案請見 p. 211

- **deserve** [dɪ`zɝv]
  值得；應得
- **effort** [`ɛfət]
  努力
- **instrumental** [ˌɪnstrə`mɛntl]
  有幫助的
- **noteworthy** [`notˌwɝði]
  值得注意的

## 1  I would like to take this opportunity to express . . .

我想藉此機會，表達……

- I would like to take this opportunity to express **my sincere thanks for your efficiency and diligence.**
  我想藉此機會，對你的效率和勤奮表達誠摯的謝意。

- On behalf of **the entire team,** I want you to know how much we appreciate **your years of service.**
  謹代表整個團隊，我想讓你知道我們有多感謝你這些年來的服務。

## 2  Noteworthy is the work you've done to . . .

值得注意的是你為了……所做的一切。

- Noteworthy is the work you've done to **increase revenues and lower costs.**
  值得注意的是你為了增加營收與降低成本所做的一切。

- Without you, we could not have **posted our best sales in nearly a decade.**
  沒有你，我們就無法繳出近十年來的最佳業績。

- Your work on **the Peterson account was instrumental in putting our company** in the black **for the year.**
  你在彼得森客戶一案的努力，促使公司能在今年獲利。

> 指「有盈餘；獲利」，相對說法為 in the red「有赤字；虧空」。

## 3 I am grateful for your willingness to . . .

我很感謝你願意……

---

• I am grateful for your willingness to **go out of your way to support your colleagues on a daily basis.**

我很感謝你願意日日不辭辛勞地支援同事。

---

 • I can't stress enough that we couldn't **have pulled** this **campaign off** without all of your efforts to help.

我怎麼強調都不為過的是,沒有你的鼎力相助,我們是無法辦好這場活動的。

> **pull sth off** 指「成功做成（困難或出乎意料的事）」。

---

## 4 As a token of our gratitude for . . . , we would like to . . .

為表達我們對……的感激之情,我們想……

---

• As a token of our gratitude for **your outstanding performance this past year,** we would like to **give you a bonus of three months' salary.**

為表達我們對你過去這一年傑出表現的感激之情,我們想給你三個月薪資的獎金。

---

 • In appreciation of **your extraordinary efforts,** we are giving you **a 5 percent raise.**

為感謝你的加倍努力,我們要幫你加薪 5%。

• I'll look for an opportunity to pay you back when **you could use some help from me in the future.**

未來你需要我幫忙的時候,我會找機會回報你的。

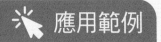

# 讚許下屬

To: brittany@cornerstoneelectronics.com
From: oliver@cornerstoneelectronics.com
Subject: Appreciation for Your Efforts

Dear Brittany,

On behalf of the entire team, I want you to know that your efforts have not gone without **notice**.[1]

Noteworthy are your sales and the work you've done to **secure**[2] several key accounts. During this past year, you **exceeded**[3] your sales target by 8 percent and helped guarantee Cornerstone Electronics a 5 percent increase in annual revenue as O&E's official supplier.

> 指「就個人來看」。

On a personal note, I am grateful for your willingness to help **newcomers**[4] get into the swing of things. Your **mentorship**,[5] communication, and coordination skills are second to none.

> 指「進入狀況；開始積極投入」。

> 指「最佳的」是較委婉的說法。

As a token of our gratitude for all you have done, we would like to offer you a 10 percent raise. You truly deserve it.

Keep up the wonderful work.

Sincerely,

Oliver Benton
Sales Manager
Cornerstone Electronics

## 中譯

親愛的布蘭妮，

謹代表整個團隊，我想讓你知道，你的努力並沒有被忽略。

值得注意的是你的業績以及你為了贏得幾個重要客戶所做的一切。過去這一年裡，你的業績目標超出 **8%**，你也協助基石電子以 O&E 公司的官方供應商之姿，守住 **5%** 的年營收成長。

就我個人而言，我很感謝你願意協助新人進入狀況。你的指導、溝通與協調的技巧不亞於任何人。

為表達我們對你所做一切的感激之情，我們想幫你加薪一成。這是你理所當得的。

繼續保持優秀的工作表現。

謹啟，

奧利佛・班頓
業務經理
基石電子

## 字彙

*1.* **notice** [ˈnotəs] *n.* 注意

*2.* **secure** [səˈkjur] *v.* 獲得

*3.* **exceed** [ɪkˈsid] *v.* 超越

*4.* **newcomer** [ˈnuˌkʌmə] *v.* 新進人員

*5.* **mentorship** [ˈmɛnˌtɔrˌʃɪp] *n.* 指導

### 延伸學習　表揚信 檢測問答句

- What achievements immediately stand out when you try to recall what the recipient has done?
  當你試著回想收件者的表現時，哪些成就是特別突出的？

- What qualities or skills does the recipient possess?
  收件者具備什麼特質或技能？

- What's it like to work alongside this particular person every day?
  每天和這個特別的人一起工作的感覺如何？

# 讚許跨部門同仁

To: keegan@cornerstoneelectronics.com
From: oliver@cornerstoneelectronics.com
Subject: Appreciation for Hard Work

Dear Keegan,

I would like to take this opportunity to express my gratitude for all the hard work you have done to keep things running smoothly. I greatly appreciate the effort you have put into **fulfilling**[1] numerous rush orders, meeting **countless**[2] client requests, and finding **cost-effective**[3] solutions whenever issues arise on the production line.

Moreover, without you, we could not have secured the O&E account. Your assistance in presenting the capabilities of our factory was instrumental in getting Cornerstone Electronics approved during the supplier **audit**.[4]

Your flexibility, professionalism, and your many meaningful contributions are **testaments**[5] to your competence and dependability.

It has been a pleasure to have you as a colleague, and I look forward to collaborating with you for many years to come.

Best Regards,

Oliver Benton
Sales Manager
Cornerstone Electronics

## 中譯

親愛的基根，

我想藉此機會，感謝你為了讓事情運作順利所做的一切努力。我非常感激你努力完成多筆急單、滿足無數客戶的需求，以及每當產線出問題時，就會找出具成本效益的解決方法。

此外，沒有你，我們就無法拿下 O&E 這個客戶。你對於展現工廠產能所給予的協助，促使基石電子得以通過供應商的審核。

你的彈性、專業及許多有意義的貢獻佐證了你的能力與可靠。

有你這個同事是我的榮幸，也期待未來很多年都能繼續與你合作。

謹致，

奧利佛・班頓
業務經理
基石電子

## 字彙

1. **fulfill** [fʊlˋfɪl] *v.* 完成

2. **countless** [ˋkaʊntləs] *adj.* 無數的

3. **cost-effective** [ˋkɔstəˋfɛktɪv]
   *adj.* 具成本效益的

4. **audit** [ˋɔdət] *n.* 審核

5. **testament** [ˋtɛstəmənt] *n.* 證明

**參考答案**

1. I would like to take this opportunity to express my sincere thanks for your hard work.

2. Noteworthy is the work you've done to prevent possible loss of data.

3. I am grateful for your willingness to go the extra mile for your colleagues.

4. As a token of our gratitude for your contributions, we would like to give you a bonus of one month's salary.

# Business Greetings
## 商務問候信

## 課前寫作練習

商務問候信怎麼寫？
請參考主題詞彙、中譯及括號內的英文提示，將下列句子翻譯成英文。

**❶** 我們想藉此機會表達對您堅定（**unwavering**）支持的誠摯謝意。

 _____

_____

主題
詞彙

- **feature** [ˋfitʃɚ]
  以……為特色
- **happiness** [ˋhæpinəs]
  幸福

- **launch** [lɔntʃ]
  推出
- **partnership**
  [ˋpɑrtnɚˌʃɪp] 夥伴關係

❷ 我們致力提供顧客第一流的貨品和服務。

(寫)

_____

_____

❸ 我很高興宣布我們最新的冬季女性運動服系列。

(寫)

_____

_____

❹ 祝您邁入新的一年有滿滿的繁榮與成功。

(寫)

_____

_____

參考答案請見 p. 219

• **prosperity** [prɑˋspɛrəti]
  繁榮
• **soar** [sɔr]
  飛騰；昂揚

• **strive** [straɪv]
  努力
• **success** [səkˋsɛs]
  成功

 TRACK 072

# 1 We would like to take this opportunity to express sincere gratitude for . . .

我們想藉此機會表達對……的誠摯謝意。

- We would like to take this opportunity to express sincere gratitude for **your continued patronage.**
  我們想藉此機會表達對您持續惠顧的誠摯謝意。

 相關寫法
- Thank you for your past interest in **our products and the chance to provide you with a quote for our latest series.**
  感謝您先前對我們產品的興趣，也謝謝您給我機會，為您提供我們最新系列的報價。

# 2 We strive to provide our customers . . .

我們致力提供顧客……

- We strive to provide our customers **quality products and exceptional service.**
  我們致力提供顧客優質的產品和出色的服務。

 相關寫法
- It was my pleasure to assist you with **your last order of the LCD panels.**
  很榮幸上次協助您訂購液晶螢幕面板。
- I welcome the chance to **give you an in-person product demo at your convenience.**
  我期盼有機會在您方便時能親自為您解說產品。

  > 指「親自的；本人的」。

# 3 I am delighted to announce our newest . . .

我很高興宣布我們最新的……

- **I am delighted to announce our newest robot vacuum.**
  我很高興宣布我們最新的掃地機器人。

- Please be the first to know that **All-Day Care is now taking pre-orders for the Luminous Weightless Foundation.**
  我們優先讓您知道全天候護膚公司已開始預接亮采無重力粉底液的訂單。

- We are a **Taiwanese manufacturer of household furniture, preparing for the launch of our latest series this summer.**
  我們是台灣的居家家具製造商,準備在今年夏天推出最新系列。

# 4 May you soar into the new year filled with . . .

祝您邁入新的一年有滿滿的……

- **May you soar into the new year filled with peace and joy.**
  祝您邁入新的一年有滿滿的平安和喜樂。

- Wishing you a wealth of success in **the coming year!**
  祝您來年事事成功!

- Wishing you **good health, good luck, and much happiness** throughout the year!
  祝您一整年身體健康、鴻運當頭、萬事如意!

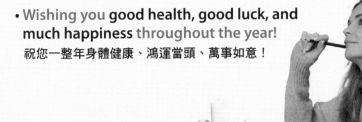

 應用範例

 TRACK 073

# 既有客戶篇

To: jlin@prizetech.com
From: anthonyhsu@illuminemanufacturing.com
Subject: Gratitude for Trust
Attached: Spec Sheet.pdf

Dear Ms. Lin:

We would like to take this opportunity to express sincere gratitude for our ongoing partnership. We appreciate the continuing **faith**[1] you have in us. It has been a pleasure helping you reach your goals this past year.

We strive to provide our customers the very best and latest technology on the market. Please be the first to know that Illumine Manufacturing is now taking pre-orders for Taiwan's first 8K LCD panels with a **pixel**[2] **density**[3] of 510 pixels per inch for truly lifelike images. Included for your **perusal**[4] are the full **specifications**[5] and data sheet, but don't take our word for it — let our on-site demonstration **convince**[6] you instead.

Wishing you a wealth of success in the coming year!

> take sb's word for it 指「相信某人的話」。

Anthony Hsu
Sales Consultant
Illumine Manufacturing

林女士您好：

我們想藉此機會表達對您我長期合作的誠摯謝意。感謝您一直以來對我們的信任。我們很高興過去這一年能協助您達成貴司的目標。

我們致力提供顧客市場上最好且最新的科技產品。我們優先讓您知道光耀製造公司已開始預接台灣第一款 8K 液晶顯示面板的訂單，此面板的像素密度為每英寸 510 像素，呈現出極逼真畫面。附上供您詳閱的完整產品規格和數據表，但不要就這樣照單全收──讓我們的現場示範來說服您吧。

祝您來年事事成功！

安東尼‧徐
銷售顧問
光耀製造公司

## 字彙

1. **faith** [feθ] *n.* 信任

2. **pixel** [ˈpɪksəl] *n.* 像素

3. **density** [ˈdɛnsəti] *n.* 密度

4. **perusal** [pəˈruzəl] *n.* 詳細閱讀

5. **specification** [ˌspɛsəfəˈkeʃən] *n.* 規格

6. **convince** [kənˈvɪns] *v.* 說服

## 延伸學習　商務問候信 寫作要點

- **Start with a complimentary or friendly remark.**
  以讚美或友善的言詞起頭。

- **State your main message.**
  表達主要的訊息。

- **Expand on the message.**
  進一步說明訊息。

- **Close with wishes for success and a mention of continued contact.**
  以祝願成功和期許能繼續聯繫作結。

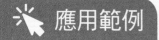 

# 潛在客戶篇

To: tamaki@jptech.com
From: anthonyhsu@illuminemanufacturing.com
Subject: Announcement of New Product
Attached: Spec Sheet.pdf

Dear Mrs. Tamaki:

Thank you for your past interest in our products and the chance to provide you with a **quote**[1] for 17" flat screen computer monitors on December 12, 2020.

I am delighted to announce our newest 8K LCD panels featuring a 120Hz refresh rate, which are **poised**[2] to **command**[3] a high percentage of Taiwan's market share, launching July 2021. We are proud of the increased price-performance ratio of these **outstanding**[4] displays, which have already impressed clients such as Sol Bank and the Cloud 9 Group of hotels.

> 指「性能價格比」，也被廣泛稱為 CP 值（**cost-performance ratio**），為一個產品根據它的價格所提供的性能高低。

I welcome the chance to **conduct**[5] an in-person product demo at your convenience.

May you soar into the new year filled with health, happiness, and prosperity.

Anthony Hsu
Sales Consultant
Illumine Manufacturing

## 中譯

玉木女士您好：

感謝您先前對我們產品的興趣，也謝謝您在 2020 年 12 月 12 日時給我機會，為您提供 17 吋平板電腦螢幕顯示器的報價。

我很高興宣布我們最新的 8K 液晶螢幕面板主打 120 赫茲的更新頻率，將於 2021 年 7 月推出，且已準備好要拿下台灣的高市占率了。我們為這些出色的顯示器所提升的性價比感到驕傲，這已讓索爾銀行和九雲飯店集團等客戶讚不絕口。

我期盼有機會在您方便時能
親自為您解說產品。

祝您邁入新的一年有滿滿的
健康、幸福以及興隆事業。

安東尼・徐
銷售顧問
光耀製造公司

## 字彙

1. **quote** [kwot] *n.* 報價

2. **poised** [pɔɪzd] *adj.* 準備好的

3. **command** [kə`mænd]
   *v.* 控制；管轄

4. **outstanding** [aʊt`stændɪŋ]
   *adj.* 出眾的

5. **conduct** [kən`dʌkt] *v.* 進行

參考答案

1. We would like to take this opportunity to express sincere gratitude for your unwavering support.

2. We strive to provide our customers the first-rated goods and services.

3. I am delighted to announce our newest winter collection of women's sportswear.

4. May you soar into the New Year filled with prosperity and success.

# Business Announcements

## 營業通知

## 課前寫作練習

營業通知怎麼寫？

請參考主題詞彙、中譯及括號內的英文提示，將下列句子翻譯成英文。

❶ 我們很高興宣布超級國際公司（Super International）將進駐密蘇里州堪薩斯城（Kansas City, Missouri）。

 _____

_____

主題
詞彙

- **accommodate**
  [əˋkɑməˌdet] 因應
- **announce** [əˋnauns]
  宣布

- **branch** [bræntʃ]
  分部
- **commemorate**
  [kəˋmɛməˌret] 慶祝

❷ 敝司於亞洲各地皆有分部，為中小企業提供企業資源計畫的服務。

(寫)

_____

_____

❸ 為因應壯大中的事業，我們決定搬遷至更大的廠址。

(寫)

_____

_____

❹ 我們的電郵和電話資訊維持不變，而我們現在新的郵寄地址是加州洛杉磯阿姆布雷斯特大道（ Armbrester Drive ）1691 號。

(寫)

_____

_____

參考答案請見 p. 227

- **effective** [ɪˋfɛktɪv]
  自……起生效的
- **mailing address**
  郵寄地址

- **operating hours**
  營業時間
- **potential** [pəˋtɛnʃəl]
  發展潛能

## 1 We are pleased to announce that (company) is coming to (location).

我們很高興宣布（公司）將進駐（地點）。

- We are pleased to announce that **Boom Burgers is coming to Sudbury, Ontario.**

  我們很高興宣布布姆漢堡將進駐安大略省薩德伯里市。

  - This is to announce the establishment of (company) in (location) for the purpose of . . .

    此信旨在宣布為了……而在（地點）建立（公司）。
  - Our newest establishment will open at (time) on (date).

    我們最新的營業據點將於（日期）（時間）開幕。

## 2 Our company, which has branches across (area), provides . . .

敝司於（地區）各地皆有分部，提供……

- Our company, which has branches across **Europe,** provides **financial advice to small- and medium-sized businesses.**

  敝司於歐洲各地皆有分部，為中小企業提供財務建議。

  - Seeing great potential in (location), we cannot wait to . . .

    我們在（地點）看到極大的發展潛能，等不及要……

## 3 To accommodate our growing business, we decided to move to . . .

為因應壯大中的事業，我們決定搬遷至……

- To accommodate our growing business, we decided to move to **a larger storefront.**

  為因應壯大中的事業，我們決定搬遷至更大的店面。

- Due to (reason), we decided to move our company from the current location to . . .

  基於（理由），我們決定將公司從現址遷至……

- We would like to inform you that on (date), our business operations will be relocated at . . .

  我們想通知您，在（日期）當天，我們的營運據點將轉移至……

## 4 Our e-mail and phone details remain unchanged, while our new mailing address is now . . .

我們的電郵和電話資訊維持不變，而我們現在新的郵寄地址是……

- Our e-mail and phone details remain unchanged, while our new mailing address is now **239 Patterson Boulevard, Houston, TX 77003, USA.**

  我們的電郵和電話資訊維持不變，而我們現在新的郵寄地址是 77003 美國德州休士頓市派特森大道 239 號。

- Our landline has changed, so please feel free to contact us at (new number).

  我們的室內電話已更改，所以歡迎透過（新號碼）與我們聯繫。

# 開業通知

To: Undisclosed Recipients
From: michaeloleary@electrify.com
Subject: Electrify at Greenwich

Greetings from the staff at Electrify,

We are pleased to announce that Electrify is coming to Greenwich. Our company, which has branches across North America, provides **consultation**[1] and **installation**[2] services for businesses looking to **outfit**[3] their **premises**[4] with solar panels.

Seeing great potential in Greenwich, we cannot wait to get **acquainted**[5] with you all in the weeks and months to come.

To commemorate joining this fantastic community, we will be holding a **BBQ**[6] at our grand opening on May 15. Come by! Our doors are open to all.

Best Regards,

Michael O'Leary
CEO
Electrify

## 中譯

來自伊萊特菲公司全體職員的問候，

我們很高興宣布伊萊特菲公司將進駐格林威治。敝司於北美各地皆有分部，提供諮詢與安裝的服務給尋求在其廠區配備太陽能板的企業。

我們在格林威治看到極大的發展潛能，等不及要在後續幾週和幾個月裡認識大家了。

為慶祝加入這個極棒的社區，我們將於 5 月
15 日盛大開幕時舉辦烤肉派對。來一趟吧！
我們竭誠歡迎大家。

謹致，

麥可・歐萊瑞
執行長
伊萊特菲公司

## 字彙

1. **consultation** [ˌkɑnsʌlˈteʃən]
   n. 諮詢

2. **installation** [ˌɪnstəˈleʃən] n. 安裝

3. **outfit** [ˈaʊtˌfɪt] n. 配備

4. **premises** [ˈprɛməsɪz] n. 廠區

5. **acquainted** [əˈkwentəd]
   adj. 熟悉的

6. **BBQ** [ˈbɑrbɪˌkju]
   n. 烤肉（barbecue 的縮寫）

## 延伸學習　營業通知 寫作要點

- **Announce the opening of your business.**
  宣布開業。

- **Add a brief message about your service or product.**
  簡短補充和服務或產品有關的訊息。

- **Invite the reader to be your customer.**
  邀請讀信者成為你的顧客。

- **Include an invitation to a special event.**
  包含特別活動的邀請。

# 遷址與營業時間變更通知

To: sarahcollins@ucc.net
From: danielsmith@greenstar.comcom
Subject: Change of Company Address and Operating Hours

Dear Sarah,

To accommodate our growing business, we decided to move to a bigger space.

Our e-mail and phone details **remain**[1] unchanged, while our new mailing address is now 181 North Mount Street, Baltimore, MD 21057, USA.

To **accompany**[2] this move, we have also **extended**[3] our operating hours. Effective April 15, Green Star will be open from 8 a.m. to 6 p.m.

We thank you for your continued support as a **valued**[4] partner. Please feel free to contact us by phone or e-mail with any questions.

Many Thanks,

Daniel Smith
President
Green Star

## 中譯

親愛的莎拉，

為因應壯大中的事業，我們決定搬遷至更大的空間。

我們的電郵和電話資訊維持不變，而我們現在新的郵寄地址是
**21057** 美國馬里蘭州巴爾的摩市北山街 **181** 號。

隨著此次的搬遷，我們也延長了營業時間。自 **4** 月 **15** 日起，
綠星公司將從早上八點營業至晚上六點。

我們感謝您身為重要夥伴的持續支持。
如有任何問題，歡迎透過電話或電子郵
件與我們聯繫。

非常感謝，

丹尼爾 · 史密斯
董事長
綠星公司

## 字彙

1. **remain** [rɪˋmen] *v.* 維持

2. **accompany** [əˋkʌmpəni] *v.* 伴隨

3. **extend** [ɪkˋstɛnd] *v.* 延長

4. **valued** [ˋvæljud] *adj.* 重要的

**參考答案**

1. We are pleased to announce that Super International is coming to Kansas City, Missouri.

2. Our company, which has branches across Asia, provides ERP services to small- and medium-sized businesses.

3. To accommodate our growing business, we decided to move to larger premises.

4. Our e-mail and phone details remain unchanged, while our new mailing address is now 1691 Armbrester Drive, Los Angeles, California.

# Letters to Potential Partners
## 業務開發信

## 課前寫作練習

業務開發信怎麼寫？
請參考主題詞彙、中譯及括號內的英文提示，將下列句子翻譯成英文。

**❶** 我想簡單介紹史密斯公司（Smitham）這間有創造力和前瞻性的公司，我們正在尋求與貴司的技術合作。

 _____

_____

主題
詞彙

- **competitive** [kəmˋpɛtətɪv]
  具競爭力的
- **enclosed** [ɪnˋklozd]
  隨信附上的

- **establishment**
  [ɪˋstæblɪʃmənt] 創立
- **outstanding**
  [autˋstændɪŋ] 卓越的

❷ 我們自 2010 年創立以來，就努力建立提供高品質電器的名聲。

(寫)

_____

_____

❸ 上城科技公司（Uptown Technologies）專司觸控筆電和平板螢幕。

(寫)

_____

_____

❹ 要了解更多有關創泰企業（Initech Corp.）的資訊，歡迎隨時寄信到 john@ initechcorp.com 聯絡我。

(寫)

_____

_____

參考答案請見 p. 235

- **premium** [ˋprimiəm]
  優質的
- **reputation** [ˌrɛpjəˋteʃən]
  名聲
- **specialize** [ˋspɛʃəˌlaɪz]
  專司
- **state-of-the-art**
  [ˋstetəvðiˋɑrt] 最先進的

## 1 I would like to briefly introduce (company), a (type of business) that is currently seeking . . .

我想簡單介紹（公司）這間（類別公司），我們正在尋求……

- I would like to briefly introduce **Urban Elements, an interior design studio** that is currently seeking **a partnership with your company.**

  我想簡單介紹城市元素這間室內設計工作室，我們正在尋求與貴司合作的可能。

 • Our mutual business associate **Sam Anderton** at **Planet Marketing** suggested that I contact you to introduce my company, **Fairlight Ltd.**

  我們在星球行銷公司的共同商業夥伴山姆‧安德頓建議我跟您聯繫，向您介紹敝司費爾萊特有限公司。

## 2 We have worked hard since our establishment in (year) to build a reputation for . . .

我們自（年分）創立以來，就努力建立……的名聲。

- We have worked hard since our establishment in **2008** to build a reputation for **providing energy-efficient appliances.**

  我們自 2008 年創立以來，就努力建立提供節能電器的名聲。

 • Within the industry, we are known for **our award-winning products and premium service.**

  在業界裡，我們以屢獲殊榮的產品和優質服務而聞名。

## 3   At (company), we specialize in . . .

（公司）專司……

> 為 computer numerical control
> 「電腦數值控制」的首字縮略語。

- **At Hollster Technology,** we specialize in **advanced 3D printing and precision CNC machining.**
  后斯特科技公司專司先進 3D 列印和精密電腦數值控制加工。

**相關寫法**
- As a worldwide leader in **relocation services, Fly Away Home** has the expertise to **make your move a resounding success.**
  身為搬家服務的全球領導者，返家公司擁有讓您搬遷一切順利的專業技術。
- Our products are manufactured in **our state-of-the-art production facilities here in Taiwan.**
  我們的產品是由台灣最先進的生產設施所製造。

---

## 4   To learn more about (company), please feel free to . . .

要了解更多有關（公司）的資訊，歡迎隨時……

- **To learn more about Bloom Technologies,** please feel free to **contact me at 02-553-7800.**
  要了解更多有關旺盛科技公司的資訊，歡迎隨時撥 02-553-7800 這支電話號碼找我。

**相關寫法**
- I am available to **speak further at your convenience** and have included my contact information **below.**
  您方便的時候我能跟您進一步詳談，我也把自己的聯絡資訊附在下面了。

# 公關公司篇

To: mhutchinson@sayitstraight.com
From: scott@mediaworks.com
Subject: Company Intro of Media Workss
Attached: Media Works_brochure.pdf

Dear Ms. Hutchinson,

I would like to briefly introduce Media Works, a young and fresh PR agency that is currently seeking more partnerships with **forward-thinking**[1] companies like Say It Straight.

> 指「企業社會責任」，一種道德或意識型態理論，主要探討企業是否有責任對社會做出貢獻。

We have worked hard since our establishment in 2005 to build a reputation for promoting corporate social responsibility to **inspire**[2] the world of commerce in innovative directions. Public relations campaigns by Media Works are regularly honored with industry awards for their creativity and impact.

Please find enclosed our current **brochure**[3] with detailed **offerings**,[4] competitive pricing, and lists of awards and clients. We also invite you to **peruse**[5] our website and consider attending one of our regular Social Change Evenings in Manhattan.

Thank you for your time, and I look forward to speaking with you in the near future.

Sincerely,

Scott Richardson
Director of Strategic Partnerships
Media Works

中譯

哈青森女士您好，

我想簡單介紹米迪亞這間非常年輕的公關經紀公司，我們正在尋求更多與直言這樣具前瞻性的公司合作的機會。

我們自 2005 年創立以來，就努力建立名聲，推廣企業社會責任以激發商場上更多的創新靈感。米迪亞所舉辦的公關活動因本身的創造力和影響力而屢獲業界殊榮。

隨函附上我們目前的手冊，內含詳細的服務介紹、具競爭力的價格，以及曾獲獎項和客戶的名單。我們也請您瀏覽我們的網站，並考慮參加我們定期在曼哈頓舉辦的「社會變革之夜」活動。

感謝您寶貴的時間，我非常期待在不久的將來能與您會談。

謹啟，

史考特‧理查森
策略合作部協理
米迪亞公關公司

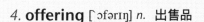

## 字彙

1. **forward-thinking** [ˈfɔrwədˈθɪŋkɪŋ] *adj.* 具前瞻性的

2. **inspire** [ɪnˈspaɪr] *v.* 激勵

3. **brochure** [broˈʃur] *n.* 小冊子

4. **offering** [ˈɔfərɪŋ] *n.* 出售品

5. **peruse** [pəˈruz] *v.* 瀏覽

 延伸學習　**業務開發信 寫作要點**

- **Arouse interest in the first paragraph.**
  在首段引起興趣。

- **Create a desire for the product or service.**
  創造對產品或服務的渴望。

- **Point out the benefits and give evidence.**
  說明優勢並提供證據。

- **Persuade the reader to take appropriate action.**
  說服讀信者採取適當的行動。

# 科技公司篇

To: kendrastevens@argomobile.com
From: robertlee@powergo.com
Subject: Greetings from PowerGo

Dear Ms. Stevens:

I am writing on the recommendation of our mutual business **associate**[1] Andrew Bird at Highwater Consultants. Andrew suggested that I contact you to introduce my company, PowerGo.

At PowerGo, we specialize in advanced consumer power solutions, with a focus on premium laptop **adapters**[2] and USB **chargers**.[3] Within the industry, we are known for the outstanding quality of our products, which are manufactured in our state-of-the-art production facilities here in Taiwan. It should come as no surprise, then, that our sales volume surpassed one million units last year and that we are on track to double that figure this year.

We understand that your company, Argo Mobile, is planning to expand its business, and we are confident that PowerGo can help you to achieve your goals.

To learn more about PowerGo, please feel free to **browse**[4] our website at www.powergo.com.tw or contact me at 02-282-4419 for more details.

We at PowerGo look forward to working with you soon.

Sincerely,

Robert Lee
Sales and Development Manager
PowerGo

## 中譯

親愛的史蒂芬斯女士：

透過我們在高水顧問公司的共同商業夥伴安德魯‧柏德的引薦，我寫了這封信。安德魯建議我跟您聯繫，向您介紹我的公司動力前進。

動力前進公司專司先進顧客電力解決方案，主打頂級筆電轉接器和 **USB** 充電器。在業界裡，我們以卓越的產品品質而聞名，我們的產品是由台灣最先進的生產設施所製造。不意外地，我們的銷量於去年突破一百萬組，而今年我們的銷量很有可能增加一倍。

我們獲悉貴司南船行動正計畫拓展業務，動力前進公司有信心能助您達成目標。

要了解更多有關動力前進公司的資訊，歡迎隨時
瀏覽我們的網站 **www.powergo.com.tw** 或撥
**02-282-4419** 聯絡我以瞭解更多訊息。

動力前進公司的全體同仁期盼很快就能與您合作。

謹啟，

羅柏特‧李
業務發展部經理
動力前進公司

## 字彙

*1.* **associate** [əˋsoʃiət] *n.* 夥伴

*2.* **adapter** [əˋdæptə] *n.* 轉接器

*3.* **charger** [ˋtʃɑrdʒə] *n.* 充電器

*4.* **browse** [brauz] *v.* 瀏覽

---

**參考答案**

1. I would like to briefly introduce Smitham, a creative and forward-thinking company that is currently seeking technical cooperation with your company.

2. We have worked hard since our establishment in 2010 to build a reputation for providing quality electrical goods.

3. At Uptown Technologies, we specialize in touch-screen laptops and flat-panel monitors.

4. To learn more about Initech Corp., please feel free to contact me at john@ initechcorp. com.

# Unit 27 Meeting Confirmation Letters
## 業務會面確認信

### 課前寫作練習

業務會面確認信怎麼寫？
請參考主題詞彙、中譯及括號內的英文提示，將下列句子翻譯成英文。

**❶** 我想提醒您我即將進行的拜訪，目的是要與您談談
艾佛芙萊斯（**EverFlex**）的最新產品。

（寫）

主題
詞彙

- **appointment** [əˈpɔɪntmənt]
  （會面的）約定
- **arrange** [əˈrendʒ]
  安排

- **confirm** [kənˈfɜm]
  確認
- **demonstration**
  [ˌdɛmənˈstreʃən] 展示

❷ 我們預定於 6 月 25 日下午兩點在您城區的辦公室會面。

(寫) _____

_____

❸ 若貴司的行銷經理能參加我們的討論就太好了。

(寫) _____

_____

❹ 若您在我們會面前有任何問題,請透過我的電話號碼 725-3321 與我聯繫。

(寫) _____

_____

參考答案請見 p. 243

- **feedback** [ˋfidˏbæk]
  回饋意見
- **remind** [rɪˋmaɪnd]
  提醒
- **request** [rɪˋkwɛst]
  要求
- **schedule** [ˋskɛdʒul]
  預定

# 1 I would like to remind you of my incoming visit to ...

我想提醒您我即將進行的拜訪，目的是要……

- I would like to remind you of my incoming visit to **review our latest products.**

  我想提醒您我即將進行的拜訪，目的是要與您談談我們的最新產品。

 • I am writing to confirm my appointment with you to discuss **the terms of our licensing agreement.**

  來信旨在確認我和您的會面，目的是要討論我們的授權協議條款。

---

# 2 We are scheduled to meet at (location) on (date) at (time).

我們預定於（日期）（時間）在（地點）會面。

- We are scheduled to meet at **the Gainesboro Conference Hall on May 25 at 3 p.m.**

  我們預定於 5 月 25 日下午三點在蓋恩斯柏勒會議廳會面。

 • The meeting is arranged for **June 20 at our branch at 315 West 23rd St. and Broadmoor Ave.**

  會議安排在 6 月 20 日，地點為我們位在西 23 街 315 號與布洛德摩爾大道的分公司。

## 3 It would be appreciated if (title) could . . .

若（職稱）能⋯⋯就太好了。

- It would be appreciated if **your sales director** could **join our meeting.**

  若貴司的業務總監能參加我們的會議就太好了。

- I'd appreciate it if you could arrange for **a video conference with your branch office in Shanghai, as we would like to keep them updated.**

  如果您能安排與您上海分公司的視訊會議就太好了，因為我們想讓他們知道最新的情況。

- If you would like to **preview any of our products before we meet,** please feel free to **visit our company website at www. everflex.com.tw.**

  如果您在我們會面前想先看看我們的任一產品，請歡迎上我們的公司網站 www.everflex.com.tw。

---

## 4 If you have any questions before we meet, please contact me via . . .

若您在我們會面前有任何問題，請透過⋯⋯與我聯繫。

- If you have any questions before we meet, please contact me via **my cell phone number: +1-434-555-6545.**

  若您在我們會面前有任何問題，請透過我的手機號碼 +1-434-555-6545 與我聯繫。

- I can be reached at **simondurwald@fivetech.com** should you have any questions or concerns prior to our meeting.

  若您在我們會面前有任何問題或疑慮，可寄信至 simondurwald@fivetech.com 聯絡我。

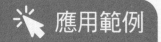

# 拜訪客戶篇

To: kelvinshaw@acat.com
From: danielhoward@crms.com
Subject: Demonstration Confirmation

Dear Ms. Kelvinshaw:

I would like to remind you of my incoming visit to **demonstrate**[1] our new customer relationship management system. Presently, we are scheduled to meet at your central branch at 1:30 p.m. on Monday, May 10.

I'd appreciate it if you could arrange for as many members of your sales team to attend the demonstration as possible, as we would welcome their questions and feedback. Additionally, please prepare a **projector**[2] in the conference room where the meeting will take place so that I may better present the system.

I can be reached at danielhoward@crms.com or via cell phone, +1-322-555-4989, should you have any questions or concerns **prior**[3] to our meeting.

I am certain you will find our newest CRM system to your **satisfaction.**[4] Thank you for your confirmation in advance, and I look forward to doing business with you.

Daniel Howard
Sales Associate
CRMS Software Co.

中譯

凱文蕭女士您好：

我想提醒您我即將進行的拜訪，目的是要展示我們新的客戶關係管理系統。目前，我們預定於 **5 月 10 日星期一下午一點半**在你們的中部分公司會面。

如果您盡可能安排多一些業務團隊成員來參加這場示範的話，就太好了，因為我們非常歡迎他們的提問與回饋意見。此外，請在會議舉行的會議室準備一台投影機，如此我才能更理想地示範這套系統。

若您在我們會面前有任何問題或疑慮，可寄信至 **danielhoward@crms.com** 或撥手機號碼 **+1-322-555-4989** 聯絡我。

我確信您將會對我們最新的客戶關係管理系統感到滿意。感謝您事先確認，期待與您有生意上的往來。

丹尼爾・霍華德
銷售專員
**CRMS 軟體公司**

字彙

1. **demonstrate** [ˈdɛmənˌstret]
   *v.* 展示

2. **projector** [prəˈdʒɛktə] *n.* 投影機

3. **prior** [praɪr] *adj.* 在前的

4. **satisfaction** [ˌsætəsˈfækʃən]
   *n.* 滿意

延伸學習　**業務會面確認信檢核表**

- Confirm all the details, including date, time, and location of the meeting.
  確認會面日期、時間與地點等所有細節。

- Tell the recipient that if the agreement does not correctly reflect their understanding, they should contact you immediately.
  告知收信人若雙方協議和其認知有出入，應立刻與你聯繫。

- Thank the recipient and tell them you look forward to the meeting.
  感謝收信人並告知對方，你對此次會面深感期待。

# 客戶參訪篇

To: blankenship@volvefloor.com
From: barkley@ghmachinery.com
Subject: Meeting Confirmation
Attached: http://maps.google.com/maps=Gramlin Heavy Machinery

Dear Mr. Blankenship:

I am writing to confirm our meeting to review my company's manufacturing **facilities**.[1] The meeting is arranged for Monday, June 12, at 9 a.m. at our factory on 238 Ave. and 138 St., No. 234. I've attached a Google Maps link to this e-mail for your convenience. If you require transportation, I can make arrangements for you upon request.

It would be appreciated if your production manager could **accompany**[2] you on the tour of the facilities. We would value his **input**[3] to **ascertain**[4] which of our models would best suit your company's needs.

If you have any questions before we meet, please contact me via my cell phone number: +1-888-555-9956.

Thank you for your time, and I look forward to meeting with you soon.

Marisa Barkley
Sales Representative
Gramlin Heavy Machinery

## 中譯

布蘭肯希普先生您好：

來信旨在確認我們要檢視敝司製造設備的會面事宜。會議安排在 **6** 月 **12** 日星期一上午九點，地點是我們位在 **238** 號大道與 **138** 街 **234** 號的工廠。我已在信中附上 **Google** 地圖連結，方便您參考。如果您需要接送，我能根據您的需求來安排。

若貴司的產線經理能陪同您參訪設備就太好了。我們相當重視他的意見，如此才得以確定我們的哪種機型最適合貴司的需求。

若您在我們會面前有任何問題，請透過我的手機號碼 **+1-888-555-9956** 與我聯繫。

感謝您的寶貴時間，期待很快與您見面。

瑪莉莎・巴克利
業務代表
格雷姆林重機械

## 字彙

*1.* **facility** [fə`sɪlətɪ] *n.* 設備

*2.* **accompany** [ə`kʌmpənɪ] *v.* 陪同

*3.* **input** [`ɪn͵pʊt] *n.* 意見

*4.* **ascertain** [͵æsɚ`ten] *v.* 確定

> **參考答案**
>
> 1. I would like to remind you of my incoming visit to review EverFlex's latest products.
> 2. We are scheduled to meet at your downtown office on June 25 at 2 p.m.
> 3. It would be appreciated if your marketing manager could join our discussion.
> 4. If you have any questions before we meet, please contact me via my phone number: 725-3321.

# Follow-Up Letters
## 業務追蹤信

## 課前寫作練習

業務追蹤信怎麼寫？
請參考主題詞彙、中譯及括號內的英文提示，將下列句子翻譯成英文。

❶ 很開心與您會面商討敝司的最新系列商品。

 ..................................................................................

..................................................................................

主題詞彙

- **brochure** [bro`ʃʊr]
  （產品）手冊
- **hear from**
  收到……的消息

- **lineup** [`laɪn,ʌp]
  （產品）系列
- **material** [mə`tɪrɪəl]
  資料

❷ 如您可能還記得的，我是生活公司（**Live Company**）的銷售總監。

(寫)

_____

_____

❸ 此趟拜訪確認了我所相信的，亦即我們升級的系統有助於促進貴司的運作。

(寫)

_____

_____

❹ 能示範如何使用此設備會非常棒。

(寫)

_____

_____

參考答案請見 p. 251

- **mission** [ˈmɪʃən]
  使命
- **recall** [rɪˈkɔl]
  記得

- **strength** [strɛŋθ]
  優勢；強項
- **suit** [sut]
  適合

## 1 I enjoyed meeting with you to discuss ...

很開心與您會面商討……

- I enjoyed meeting with you to discuss **our new software offerings.**

  很開心與您會面商討敝司的新軟體產品。

- It was a pleasure meeting you at **the Beauty Fair** on **November 3.**

  很開心 11 月 3 日能在美容展上認識您。

- I very much appreciate your meeting with me to **go over the service that my company could provide.**

  非常感謝您與我會面，讓我說明敝司所能提供的服務。

## 2 As you may recall, I am (title) at (company).

如您可能還記得的，我是（公司）的（職稱）。

- As you may recall, I am **the head of Purchasing at Mercantile.**

  如您可能還記得的，我是麥肯泰爾公司的採購主管。

- I am currently working as **the marketing manager** at **Safe-Home Security System.**

  我目前是安家安全系統的行銷經理。

## 3 The visit confirmed my belief that our (product/service) would help to . . .

此趟拜訪確認了我所相信的，亦即我們的（產品／服務）有助於⋯⋯

- The visit confirmed my belief that our **3D printers would help to improve your prototyping process.**

  此趟拜訪確認了我所相信的，亦即我們的 3D 列印機有助於改善貴司的模型製程。

- I am confident that our products are perfectly suited for your salon business.

  我有信心我們的產品絕對適合您的美髮事業。

- **Simulacrum** takes pride in producing **the best reference monitors** in the business.

  西謬拉肯公司相當自豪於出產業界最佳的參考監視器。

---

## 4 It would be wonderful to . . .

能⋯⋯會非常棒。

- It would be wonderful to **go over the details with you in person.**

  能親自向您說明細節會非常棒。

- I'm more than happy to **send you some samples of our paints.**

  我相當樂意寄一些我們的塗料樣品給您。

- I'd be delighted to **do a quick review of the proposal and answer any pending questions.**

  我很樂意快速說明這份提案，並回答任何未解決的問題。

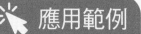 

# 會後聯繫追蹤信

To: cecilia@avissena.com
From: ross@eliteeducation.com
Subject: Follow-Up on Last Meeting

Hi Cecilia,

I enjoyed meeting with you to discuss McGraw's plan to **outsource**[1] its training program as well as its current challenges. The visit confirmed my belief that our courses would very much help to **reinforce**[2] and build on the strengths of your organization.

I'm more than happy to send you some extra materials so that you can get a better sense of what we offer. If you'd like, we'd also be willing to do a sample lesson. Our company takes great pride in our trainers, courses, and **central**[3] mission. Nothing matters to us more than providing a quality service to our clients.

> take pride in 指「對……感到自豪」，相當於 be proud of。

Thank you for taking the time to consider Elite Education. I hope to hear from you soon so we can set start dates for the courses.

Sincerely,

Ross Gilbert
Training Specialist
Elite Education

## 中譯

西西莉亞您好，

很開心與您會面商討麥克葛羅公司委外培訓課程的計畫，以及貴司當前的挑戰。此趟拜訪確認了我所相信的，亦即我們的課程非常有助於強化和建構貴組織的優勢。

我相當樂意寄一些額外的資料給您，這樣您就能更了解我們所提供的服務。您願意的話，我們也很樂意進行一堂示範課程。敝司相當自豪於自家的講師、課程及主要使命。對我們而言，最重要的是為客戶提供優質的服務。

感謝您撥冗考慮菁英教育公司。期待很快就收到您的回覆，
這樣我們就能訂定開課的日期了。

謹啟，

羅斯・吉爾伯特
培訓專員
菁英教育公司

## 字彙

1. **outsource** [ˋautˏsɔrs] v. 委外；外包

2. **reinforce** [ˏriənˋfɔrs] v. 強化

3. **central** [ˋsɛntrəl] adj. 主要的

## 延伸學習　商務邀請函 組成元素

- Establish a context.
  建構出前情脈絡。

- Remind the reader of your identity.
  提醒讀信者你的身分。

- State the advantages of your product or service.
  闡明公司產品或服務的優勢。

- Tailor your pitch to make it compelling.
  客製化你的話術，使其具有說服力。

- Express desire to work with the reader.
  表達與讀信者合作的渴望。

# 潛在客戶追蹤信

To: walter@bvzoonm.com
From: william@swiftmotors.com
Subject: Follow-Up on Last Meeting
Attached: Product.pdf

Hi Walter,

It was a pleasure meeting you at E-tech on November 10. As you may recall, I am the head of sales at Swift Motors. Since you asked for more information about Swift's electric motor lineup, I have attached some brochures for you to peruse.

I am confident that our products are perfectly suited for Newton's high-performance electric sports cars. As you will see for yourself, no other manufacturer's engines come close to the total power **output**[1] or max **torque**[2] of our **top-of-the-line**[3] **synchronous**[4] electric motors.

> bring sth to the table 指「貢獻（提議或想法）」。

It would be wonderful to discuss what Swift can bring to the table in greater depth. I'd be happy to talk over the phone or sit down with you in person. Let me know if you have any questions. I'm at your disposal.

Best Regards,

> at sb's disposal 指「任憑差遣、吩咐」。

William Frazier
Head of Sales
Swift Motors

## 中譯

華特您好，

很開心 11 月 10 日能在電子科技展上認識您。如您可能還記得的，我是捷迅引擎公司的業務主管。由於您要求更多有關捷迅電動馬達系列的資訊，所以我附上一些手冊供您閱覽。

我有信心我們的產品絕對適合紐頓的高效能電動跑車。如您將親眼所見，沒有其它製造商的引擎可媲美敝司頂級同步電動馬達的總馬力輸出或最大扭力值。

能更深入討論捷迅所能提供的會非常棒。我很樂意在電話中或坐下來親自與您洽談。如有任何問題，請讓我知道。我隨時願意為您服務。

謹致，

威廉‧佛瑞茲爾
業務主任
捷迅引擎公司

## 字彙

1. **output** [ˋaut͵put] *n.* 輸出

2. **torque** [tɔrk] *n.* 扭力

3. **top-of-the-line** [ˋtɑp͵əvðəˋlaɪn] *adj* 頂級的

4. **synchronous** [ˋsɪŋkrənəs] *adj.* 同步的

---

参考答案

1. I enjoyed meeting with you to discuss the latest lineup of our company.

2. As you may recall, I am the director of Sales at Live Company.

3. The visit confirmed my belief that our upgraded system would help to facilitate the operation of your company.

4. It would be wonderful to demonstrate how to use this equipment.

# Trades and Transactions
# 貿易往來篇

詢價信

報價信

訂購信

出貨緊急
情況通知

出貨通知

請款信

催貨通知

催款信

客訴信

# Unit 29

# Price Inquiries
## 詢價信

## 課前寫作練習

詢價信怎麼寫？
請參考主題詞彙、中譯及括號內的英文提示，將下列句子翻譯成英文。

**❶** 友善飛行公司（**Friendly Flyers**）告知我們貴司是一系列舒適旅行用枕的製造商。

 _____

_____

主題
詞彙

- **discount** [ˋdɪˏskaʊnt]
  折扣
- **material** [məˋtɪrɪəl]
  材料

- **obliged** [əˋblaɪdʒd]
  感激的
- **product line**
  產品線

❷ 我們相信貴司的產品在台灣市場大有可為。

㊢ _____

_____

❸ 如果你們能寄送更多貴司各產品線的詳細資訊，我
們將不勝感激。

㊢ _____

_____

❹ 請說明購買超過兩萬件的數量時所能享有的折扣。

㊢ _____

_____

參考答案請見 p. 259

- **quantity** [ˋkwɑntəti]
  數量
- **quote** [kwot]
  報價

- **sample** [ˋsæmpəl]
  樣品
- **terms of payment**
  付款條件

**1** **(Company) has informed us that you are . . .**

（公司）告知我們貴司是……

- **Spett Textiles has informed us that you are manufacturers of handmade gloves in a variety of natural fibers.**

  史裴特紡織公司告知我們貴司是各類天然纖維手工手套的製造商。

- I am writing to inquire about **the air purifiers that your factory produces.**

  來信旨在詢問關於貴廠所生產的空氣濾淨器。

- I am writing in connection to your advertisement for **the latest SUV lineup.**

  來信主要和貴司最新休旅車系列的廣告有關。

---

**2** **We believe there is a promising market in (location) for your products.**

我們相信貴司的產品在（地點）市場大有可為。

- We believe there is a promising market in **West Europe** for your products.

  我們相信貴司的產品在西歐市場大有可為。

- There is a steady demand in **the northwest region for high-quality headsets.**

  高品質耳機在西北區有穩定的需求量。

# 3 We would be much obliged if you could send us more details about . . .

如果你們能寄送更多……的詳細資訊，我們將不勝感激。

- We would be much obliged if you could send us more details about **your office dehumidifiers.**

  如果你們能寄送更多貴司辦公除濕機的詳細資訊，我們將不勝感激。

 • We would be grateful if you could supply **samples of material from which the sports jackets are made.**

如蒙提供製作這些運動夾克的材質樣本，我們將不勝感激。

---

# 4 Please state the discounts available for purchases of quantities of . . .

請說明購買……的數量時所能享有的折扣。

- Please state the discounts available for purchases of quantities of **more than 50,000 pieces.**

  請說明購買超過五萬件的數量時所能享有的折扣。

 • Please specify the discounts allowed on purchases of quantities of **15,000 units.**

請說明購買一萬五千組的數量所能給的折扣。

- Could you tell us what discounts you offer on purchases of quantities of **10,000 units?**

  你們能告訴我們訂購一萬組的數量
  會提供什麼樣的折扣？

# 產品詢價

To: twang@techmac.com

From: pp@ultratech.com

Subject: Product Inquiry: Smartphone and Tablet PC Cases and Covers

Dear Mr. Wang:

JetSet Technology Products has informed us that you are manufacturers of premium smartphone and tablet PC cases and covers.

We are UltraTech, a wholesaler of **high-end**[1] electronic goods and **accessories**,[2] and we believe there is a **promising**[3] market in North America for your products.

In order for us to learn more about the items you offer, we would be much obliged if you could send us more details about your various product lines. Please include sizes, colors, and prices, along with samples of the different models you manufacture and the materials used in these goods.

In addition, please state your terms of payment and the discounts available for purchases of quantities of more than 500 of a specific item. Kindly note that all prices quoted should include the cost of shipping to our warehouse.

We look forward to working with you, and your **prompt**[4] reply is appreciated.

Sincerely,

Pam Pricewater
UltraTech

## 中譯

王先生您好：

傑賽特科技產品公司告知我們貴司是優質智慧型手機和平板電腦保護殼與保護套的製造商。

我們是奧特拉科技公司，高檔電子產品與配件的批發商，我們相信貴司的產品在北美市場大有可為。

為了讓我們更了解你們所提供的產品，如果你們能寄送更多貴司各產品線的詳細資訊，我們將不勝感激。請包括尺寸、顏色和價格等資訊，還有貴司不同型號的製造樣本，以及這些產品所使用的材料。

此外，請說明貴司的付款條件與購買單項特定商品超過五百件的量時所能享有的折扣。提醒您，所有報價應包含運送至我們倉庫的費用。

我們期盼與您合作，也感謝您的及時答覆。

謹啟，

潘·普萊斯瓦特
奧特拉科技公司

## 字彙

1. **high-end** [`haɪˋɛnd] *adj.* 高檔的

2. **accessory** [əkˋsɛsəri] *n.* 配件

3. **promising** [`pramɪsɪŋ] *adj.* 大有可為的

4. **prompt** [prampt] *adj.* 及時的

延伸學習 **詢價信 寫作檢核表**

- Say how you obtained the company's information.
  表明你如何得知該公司的資訊。

- Introduce your company and describe your field of activity.
  介紹自家公司並描述業務範圍。

- State clearly what you want: a catalog, samples, terms of payments, etc.
  清楚說明你要的東西：型錄、樣本、付款條件等。

---

參考答案

1. Friendly Flyers has informed us that you are manufacturers of a range of cozy travel pillows.

2. We believe there is a promising market in Taiwan for your products.

3. We would be much obliged if you could send us more details about your product lines.

4. Please state the discounts available for purchases of quantities of more than 20,000 units.

# Quotation Letters
## 報價信

## 課前寫作練習

報價信怎麼寫？
請參考主題詞彙、中譯及括號內的英文提示，將下列句子翻譯成英文。

❶ 感謝您於 3 月 6 日來信詢問更多關於我們裝飾服務的訊息。

寫 _____

_____

主題
詞彙

- **carriage/freight** [ˋkɛrɪdʒ] [fret]
  運費
- **discount** [ˋdɪˏskaʊnt]
  折扣

- **insurance** [ɪnˋʃʊrəns]
  保險
- **item** [ˋaɪtəm]
  品項

❷ 如您所要求的,我們附上您感興趣的五項產品的價目表。

(寫) _____

_____

❸ 開放科技公司(OpenTech)以其高品質的軟體產品著名。

(寫) _____

_____

❹ 我們也想藉此機會告知您我們買一送一的好康。

(寫) _____

_____

參考答案請見 p. 265

- **packing** [ˋpækɪŋ]
  包裝
- **price list**
  價目表
- **quantity** [ˋkwɑntətɪ]
  (訂購)數量
- **unit price** [ˋjunət]
  單價

## 1 Thank you for your e-mail inquiry on (date) requesting further information about ...

感謝您於（日期）來信詢問更多關於……的訊息。

- Thank you for your e-mail inquiry on **February 15** requesting further information about **the prices of our home appliances.**

  感謝您於 2 月 15 日來信詢問更多關於敝司家電產品價格的訊息。

 • Thank you for your inquiry dated **February 20** regarding **our electronic products.**

  感謝您於 2 月 20 日來信詢問我們的電子產品。

## 2 As requested, we are enclosing a price list for ...

如您所要求的，我們附上……的價目表。

- As requested, we are enclosing a price list for **all the products you've expressed interest in.**

  如您所要求的，我們附上所有您感興趣產品的價目表。

 • Please find enclosed a quotation form listing **all the products you are interested in.**

  隨函附上一份所有您感興趣產品的報價清單。

# 3 (Company) is renowned for . . .

（公司）以 …… 著名。

- **ScienceDirect** is renowned for **its affordable, reliable, and easy-to-use software products.**

  科學直通公司以其價格實惠、可靠且簡易使用的軟體產品著名。

 • Here at **ArtDesign,** we pride ourselves on **strict selection of materials and sound workmanship.**

藝術設計公司的全體同仁以嚴選素材和上乘做工感到自豪。

---

# 4 We'd also like to take this opportunity to inform you of . . .

我們也想藉此機會告知您……

- We'd also like to take this opportunity to inform you of **our new home-delivery service.**

  我們也想藉此機會告知您我們新的運送到府服務。

 • We feel that you may also be interested in **our other products.**

我們覺得您可能也會對我們的其他產品感興趣。

• We trust you will find our quotation satisfactory and look forward to **receiving your order.**

我們相信您會滿意我們的報價，
也期盼收到您的訂單。

# 英文報價信

---

To: flanders@FEW.com
From: john.lee@bblastproducts.com
Subject: Price Quotation
Attached: price list.pdf

---

Dear Ms. Flanders:

Thank you for your e-mail inquiry on February 18 requesting further information about the prices of our award-winning cases and accessories for **portable**[1] **devices.**[2]

As requested, we are enclosing a price list for all the products you've expressed interest in. As one of the industry's leading manufacturers of stylish, lightweight cases and accessories, Bblast is renowned for its well-designed, reliable, and **durable**[3] products.

We'd also like to take this opportunity to inform you of our new app development service. Bblast now designs and develops apps for both iOS and Android devices. Details of our new Bblast App Development service can be found on our website (www.bblast.com).

Once again, thank you for your interest in Bblast. Please e-mail or call me at any time if you need any **additional**[4] assistance.

Sincerely,

John Lee
Sales and Marketing Director
Bblast Products

## 中譯

佛蘭德斯女士您好：

感謝您於 **2** 月 **18** 日來信詢問更多關於我們獲獎無數的可攜式裝置保護殼與配件的價格訊息。

如您所要求的，我們附上所有您感興趣產品的價目表。做為時尚輕巧外殼與配件的業界領導製造商，畢布萊斯特公司以設計優良且可靠耐用的產品著名。

我們也想藉此機會告知您我們新的應用程式開發服務。畢布萊斯特目前設計並開發 **iOS** 與 **Android** 裝置的應用程式。新的畢布萊斯特應用程式開發服務的詳情可在我們 **www.bblast.com** 的網站上找到。

再次感謝您對畢布萊斯特公司的興趣。如果您需要額外的協助，請隨時來信或致電給我。

謹啟，

約翰·李
營銷總監
畢布萊斯特產品

## 字彙

*1.* **portable** [ˈpɔrtəbəl] *adj.* 便於攜帶的

*2.* **device** [dɪˈvaɪs] *n.* 裝置；儀器

*3.* **durable** [ˈdjʊrəbəl] *adj.* 耐用的

*4.* **additional** [əˈdɪʃən] *adj.* 額外的

### 延伸學習 報價信 寫作要點

- **An expression of appreciation for the inquiry**
  對詢價表達感謝

- **An indication of what the prices cover, e.g. packing, carriage, insurance**
  表明價錢的涵蓋範圍，例如：包裝、運費、保險

- **Information regarding date of delivery**
  關於運送日期的資訊

參考答案

1. Thank you for your e-mail inquiry on March 6 requesting further information about our decoration services.

2. As requested, we are enclosing a price list for the five products you are interested in.

3. OpenTech is renowned for its high-quality software products.

4. We'd also like to take this opportunity to inform you of our buy-one-get-one-free deal.

# Unit 31 Letters of Order
## 訂購信

## 課前寫作練習

訂購信怎麼寫？
請參考主題詞彙和中譯，將下列句子翻譯成英文。

❶ 感謝您於 2021 年 5 月 12 日寄來的報價單。

_____

_____

主題詞彙

- **consignee** [kənˋsaɪni]
  收貨人
- **delivery notice**
  送件通知

- **inventory** [ˋɪnvənˌtɔri]
  存貨
- **place an order**
  下單

❷ 承上述報價單，隨函附上我們的初始訂單。

(寫)

_____

_____

❸ 我們想訂購貴司的無線吸塵器。

(寫)

_____

_____

❹ 請確定你們能在 2021 年 5 月 25 日前到貨。

(寫)

_____

_____

參考答案請見 p. 271

- **quotation** [kwoˋteʃən]
  報價單
- **shipment** [ˋʃɪpmənt]
  裝運
- **shipper** [ˋʃɪpɚ]
  承運人
- **stock** [stɑk]
  庫存

# 重點句型

## 1 Thank you for your quotation of (date).

感謝您於（日期）寄來的報價單。

- **Thank you for your quotation of March 5, 2021.**
  感謝您於 2021 年 3 月 5 日寄來的報價單。

相關
寫法
- **We have received your quotation of August 15 and enclose our order form.**
  我們已收到您 8 月 15 日寄來的報價單，隨函則附上我們的訂單。

- **With reference to you quotation of April 21, we'd like to place an order for air conditioner JB213.**
  根據您 4 月 21 日寄來的報價單，我們想訂購 JB213 型號的冷氣機。

## 2 Following up on this, please find enclosed . . .

承上述報價單，隨函附上……

- **Following up on this, please find enclosed our official order form.**
  承上述報價單，隨函附上我們的正式訂單。

相關
寫法
- **Our order number 237 for four of the units is enclosed.**
  我們編號 237 號的四款裝置訂單已附上。

## 3 We would like to place an order for . . .

我們想訂購……

- **We would like to place an order for your G25 surveillance cameras.**
  我們想訂購貴司的 G25 型監視攝影機。

- **We would like to purchase 5,000 electric mixers, all in the color red.**
  我們想訂購五千組電動攪拌器，全都紅色的。
- **We'd like to start with 10,000 units of your EO190 panels.**
  我們想從一萬組貴司的 EO190 面板開始訂購。

---

## 4 Please confirm that you will be able to deliver by (date).

請確定你們能在（日期）前到貨。

- **Please confirm that you will be able to deliver by March 28, 2021.**
  請確定你們能在 2021 年 3 月 28 日前到貨。

- **We would like the delivery to be done within 14 days of the order date.**
  我們想在訂購日期起的十四天內收到貨品。
- **Prompt delivery would be appreciated as the goods are needed urgently.**
  煩請盡速交貨，因為我們急需這批貨品。

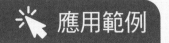 應用範例

# 英文訂購信

To: ramsey@addon.com
From: joan_acuff@ctwi.com
Subject: Placing an Order
Attached: Official Order.pdf

Dear Mr. Ramsey:

Thank you for your quotation of March 5.

Following up on this, please find enclosed our official order form. We would like to place an order for the following items: 1,000 mobile battery **packs**[1] (model AO4242) and 500 tablet **styluses**[2] (model AO3836). To confirm, the prices for these items are US$10 per unit for the mobile battery packs and US$3 per unit for the styluses.

Please **acknowledge**[3] **receipt**[4] of this order and confirm that you will be able to deliver by March 28.

Sincerely,

Joan Acuff
CompuTech Wholesalers, Inc.

## 中譯

藍姆西先生您好：

感謝您於 3 月 5 日寄來的報價單。

承上述報價單，隨函附上我們的正式訂單。我們想訂購以下物品：一千件行動電池組（型號 AO4242）和五百支平板電腦觸控筆（型號 AO3836）。向您確認，這些品項的價格為行動電池組單價十美元、觸控筆單價三美元。

請在收到訂單時告知我們，並請確定你們能在 3 月 28 日前到貨。

謹啟，

瓊恩‧艾可芙
電腦科技批發有限公司

## 字彙

1. **pack** [pæk] *n.* 一組；一套

2. **stylus** [ˈstaɪləs] *n.* 觸控筆

3. **acknowledge** [əkˈnɑlɪdʒ]
   *v.* 告知收到（信件等）

4. **receipt** [rɪˈsit] *n.* 收到；接獲

### 延伸學習　訂購信 包含要點

- an accurate and full description of goods required
  明確且完整敘述所需的貨品

- catalog numbers, quantities, and prices
  型錄編號、數量和價格

- delivery requirements such as place, date, and mode of transport
  地點、日期和運送方式等的運送需求

---

參考答案

1. Thank you for your quotation of May 12, 2021.

2. Following up on this, please find enclosed our initial order form.

3. We would like to place an order for your cordless vacuum cleaners.

4. Please confirm that you will be able to deliver by May 25, 2021.

# Shipping Emergency Notifications

## 出貨緊急情況通知

### 課前寫作練習

出貨緊急情況通知怎麼寫？
請參考主題詞彙和中譯，將下列句子翻譯成英文。

❶ 特此致函通知您所訂購貨品的數量可能有誤。

_____

_____

主題詞彙

- **completion** [kəm`pliʃən]
  完成
- **delay** [dɪ`le]
  延誤

- **error** [`ɛrɚ]
  錯誤
- **extension** [ɪk`stɛnʃən]
  延期

❷ 若您能盡速回覆，我們會非常感激，因為我們想盡快處理您的訂單。

(寫)
_____

_____

❸ 由於不可預見的情況，我們很抱歉地向您報告貴司訂單的完成時間將會延遲。

(寫)
_____

_____

❹ 對於可能對您造成的任何不便，我們感到抱歉。

(寫)
_____

_____

參考答案請見 p. 279

- **inconvenience** [ˌɪnkən`vinjəns]
  不便
- **inform/notify** [ɪn`fɔrm] [`notəˌfaɪ]
  通知
- **processing** [`prɑˌsɛsɪŋ]
  處理
- **resolve** [rɪ`zɑlv]
  解決

## 1 This letter is to inform you of the possible mistakes in . . .

特此致函通知您……可能有誤。

- **This letter is to inform you of the possible mistakes in the amounts stipulated in your purchase order.**
  特此致函通知您訂單內載明的數量可能有誤。

 相關 句型

- **Your confirmation letter dated . . . shows inconsistency in the amounts:**
  在您（日期）的確認信中，顯示的數量並不一致：

- **It is our duty to alert you to an impending delay in the completion of your order as a result of . . .**
  我們有責任提醒您，貴司訂單的完成時間將會延遲，這是由於……

## 2 Your prompt reply would be highly appreciated as . . .

若您能盡速回覆，我們會非常感激，因為……

- **Your prompt reply would be highly appreciated as swift authorization ensures that we may resolve the issue as soon as possible.**
  若您能盡速回覆，我們會非常感激，因為您的迅速授權可確保我們能盡快解決這個問題。

 相關 句型

- **Kindly reply at your earliest convenience as we wish to . . .**
  煩請盡早回覆，因為我們希望……

## 3 Due to . . ., we regret to report that . . .

由於……，我們很抱歉地向您報告……

- Due to **new legislation in our material supplier's country restricting export volume,** we regret to report that **our supply of serviceable aluminum is below required levels.**

  由於我們原料供應商的國家新立法限制出口量，我們很抱歉地向您報告，我們可供應的合用鋁料達不到需求的量。

- We are sorry to inform you that . . .

  我們很抱歉要通知您……

- Unfortunately, there has been a processing error in . . .

  很遺憾地，……流程出現錯誤。

---

## 4 We apologize for any inconvenience . . .

對於……的任何不便，我們感到抱歉。

- We apologize for any inconvenience **this may cause.**

  對於可能造成的任何不便，我們感到抱歉。

- We are truly sorry for any problems this delay may have . . .

  對此延誤可能……的任何問題，我們深感抱歉。

- Please accept our apology for any trouble . . .

  對於……的任何麻煩，請接受我們的道歉。

# 訂單數量錯誤

To: klein@abctech.com
From: timmons@kalamatov.com
Subject: Wrong Order Amount

Dear Mrs. Klein,

This letter is to inform you of the possible mistakes in the amounts **stipulated**[1] in your purchase order. We have acknowledged the receipt of your purchase order dated November 11. The details are as follows:

- Widecom chip - 6,000pcs
- digital signal processor - 5,000pcs
- SG battery - 3,000pcs

However, your confirmation letter dated November 13 shows **inconsistency**[2] in the amounts: Widecom chip 1,000 pieces short, DSP 2,000 pieces short, and SG battery 2,000 pieces more.

Your **prompt**[3] reply would be highly appreciated as swift **authorization**[4] ensures that we may **initiate**[5] **component**[6] preparation and stay within our agreed production schedule.

Sincerely,

Serge Timmons
Head of Sales
Kalamatov Manufacturing

## 中譯

克萊恩女士您好，

特此致函通知您訂單內載明的數量可能有誤。我們已確認在 11 月 11 日收到您的訂單。細節如下：

- **Widecom** 晶片——六千片
- 數位訊號處理器——五千個
- **SG** 電池——三千個

然而，您 11 月 13 日的確認信所顯示的數量並不一致：**Widecom** 晶片短少一千片，數位訊號處理器短少二千個，**SG** 電池則多出二千個。

若您能盡速回覆，我們會非常感激，因為您的迅速授權可確保我們能開始準備零件，並維持談定的生產時程。

謹啟，

賽吉・提蒙斯
業務部經理
卡拉馬多夫製造公司

## 字彙

*1.* **stipulate** [ˈstɪpjəˌlet] *v.* 載明

*2.* **inconsistency** [ˌɪnkənˈsɪstənsi] *n.* 不一致

*3.* **prompt** [prɑmpt] *adj.* 迅速的

*4.* **authorization** [ˌɔθərəˈzeʃən] *n.* 授權

*5.* **initiate** [ɪˈnɪʃɪˌet] *v.* 開始

*6.* **component** [kəmˈponənt] *n.* 零件

## 延伸學習　出貨緊急情況通知 寫作要點

- Acknowledge receipt of the order.
  告知已收到訂單。

- Explain what is wrong with the order.
  解釋訂單出了什麼問題。

- Clarify what action you want the buyer to take.
  說明你希望買方所採取的行動。

- Close by requesting that your reader address the problem immediately.
  以請求讀信者盡快處理問題作結。

# 原料供應短缺

To: chang@symtech.com
From: nakamura@sibaautomotives.com
Subject: Short on Materials

Dear Mr. Chang,

Due to new **legislation**[1] in our material supplier's country **restricting**[2] export **volume**,[3] we regret to report that our supply of **serviceable**[4] aluminum is below required levels. Thus, we are sorry to inform you that your order may not be completed by the requested deadline. We have found an alternative supplier to **supplement**[5] the aluminum and are currently **expediting**[6] production.

We would like to request a deadline extension of an additional week to ensure that the **merchandise**[7] produced meets our standard of quality. We apologize for any inconvenience this may cause. We look forward to hearing your opinions on this matter.

Sincerely,

Jim Nakamura
Chief Production Manager
Siba Automotives

## 中譯

張先生您好，

由於我們原料供應商的國家新立法限制出口量，我們很抱歉地向您報告，我們可供應的合用鋁料達不到需求的量。因此，很抱歉要通知您，您的訂單可能無法在要求的期限內完成。我們已找到替代供應商來補足鋁料，現正加速生產。

我們欲請求截止期限延長一週以確保所生產的貨物符合我們的品質標準。對於可能造成的任何不便，我們感到抱歉。我們期待得知您對此事的看法。

謹啟，

吉姆・中村
生產部經理
錫巴汽車公司

## 字彙

1. **legislation** [ˌlɛdʒəˋsleʃən] *n.* 立法

2. **restrict** [rɪˋstrɪkt] *v.* 限制

3. **volume** [ˋvɑljum] *n.* 數量

4. **serviceable** [ˋsɝvəsəbəl]
   *adj.* 合用的

5. **supplement** [ˋsʌpləˌmɛnt] *v.* 補充

6. **expedite** [ˋɛkspəˌdaɪt] *v.* 加快

7. **merchandise** [ˋmɝtʃənˌdaɪs]
   *n.* 貨物

---

**參考答案**

1. This letter is to inform you of the possible mistakes in the amounts of the goods your ordered.

2. Your prompt reply would be highly appreciated as we would like to process your order as soon as possible.

3. Due to unforeseen circumstances, we regret to report that there will be an impending delay in the completion of your order.

4. We apologize for any inconvenience this may have caused you.

## Unit 33 Shipment Notifications

# Shipment Notifications
## 出貨通知

### 課前寫作練習

出貨通知怎麼寫？
請參考主題詞彙、中譯及括號內的英文提示，將下列句子翻譯成英文。

**❶ 我們很高興向您確認，您於 4 月 15 日訂購的機械零件已包裝好並準備進行運送。**

_____

_____

**主題詞彙**

- **carrier** [ˈkɛrɪə]
  運送人
- **consignment**
  [kənˈsaɪnmənt] 寄送的貨品
- **forwarding agent**
  貨運承攬商
- **logistics** [ləˈdʒɪstɪks]
  物流

❷ 物品將由洲際物流中心（Intercontinental
Logistics）運送，應該會在 5 月 20 日前送抵貴司。

寫

_____

_____

❸ 寄送貨品包含五十箱的機械零件，每箱約重二十公
斤。

寫

_____

_____

❹ 您可放心，洲際物流中心將妥善處理您的訂單。

寫

_____

_____

參考答案請見 p. 285

- **pack** [pæk]
  包裝
- **packaging** [ˋpækɪdʒɪŋ]
  包裝材料

- **warehouse** [ˋwɛrˌhaʊs]
  倉庫
- **weigh** [we]
  有……重量

**1** **We are pleased to confirm that (item) that you ordered on (date) is packed and ready for shipping.**

我們很高興向您確認,您於(日期)訂購的(品項)已包裝好並準備進行運送。

- **We are pleased to confirm that the Olivetti KX word processors that you ordered on October 15 are packed and ready for shipping.**

  我們很高興向您確認,您於 10 月 15 日訂購的奧利維緹 KX 文字處理器已包裝好並準備進行運送。

 · **The mohair rugs you ordered on January 14 have been packed in three waterproof cases and are ready for dispatch.**

您於 1 月 14 日訂購的馬海毛地毯已包裝在三個防水箱子裡,並準備進行運送。

---

**2** **The items will be sent by (shipping company) and should arrive at your company by (date).**

物品將由(貨運公司)運送,應該會在(日期)前送抵貴司。

- **The items will be sent by Top Freight Logistics and should arrive at your company by July 21.**

  物品將由頂級貨運物流公司運送,應該會在 7 月 21 日前送抵貴司。

 · **Arrangements for shipment have been made with Atlantic Express, and the items should reach you by June 9.**

亞特蘭大快遞公司已安排裝運,貨品應該會在 6 月 9 日前送抵貴司。

## 3 The consignment consists of . . .

寄送貨品包含 ⋯⋯

- **The consignment consists of 10 cases of A2 quality printing paper, with each case weighing about 15 kilograms.**

  寄送貨品包含十箱的 A2 優質列印紙，每箱約重十五公斤。

- The items you ordered have been packed in **15 cases, with each case weighing about 30 kilograms.**

  您訂購的物品已包裝成十五箱，每箱約重三十公斤。

- The consignment awaits collection at our warehouse and consists of **six cases, each weighing about 10 kilograms.**

  寄送貨品在我們的倉庫靜候提取，總共六箱，每箱約重十公斤。

---

## 4 You can be assured that (shipping company) will take good care of your order.

您可放心，（貨運公司）將妥善處理您的訂單。

- **You can be assured that Lynden Transport will take good care of your order.**

  您可放心，林登運輸公司將妥善處理您的訂單。

- We feel sure you will find the consignments support our claim to **excellent quality.**

  我們確信您會發現寄送貨品符合
  我們對絕佳品質的要求。

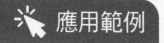

# 出貨通知

To: hillcrest@pgtechnology.com
From: paul.lee@foresite.com
Subject: Shipping Details Confirmation

Dear Ms. Hillcrest:

> laser line level「雷射水平儀」是一種發射紅色雷射光以測量水平及鉛直的工具。

We are pleased to confirm that the Foresite laser line levels that you ordered on April 28 are packed and ready for shipping.

> 指「到岸價格」，為 cost, insurance, and freight「成本、保險加運費」的簡寫，表示貨物運至買方國目的港前的全數費用和風險，由賣方負擔。

The items will be sent, CIF Keelung, by Atlas Logistics, our forwarding agent, and should arrive at your company by May 9.

The consignment **consists**[1] of 20 cases of Foresite laser line levels, with each case weighing about 25 kilograms. It will **await**[2] **collection**[3] at our warehouse.

You can be assured that Atlas Logistics will take good care of your order. We are also sure that you will find the Foresite laser line levels to be of excellent quality.

Looking forward to serving you again in the near future.

Sincerely,

Paul Lee
Foresite Precision Tools and Technology

## 中譯

希爾克瑞斯特女士您好：

我們很高興向您確認，您於 4 月 28 日訂購的佛瑞賽特雷射水平儀已包裝好並準備進行運送。

物品將由我們的貨運承攬商亞特拉斯物流公司以到岸價格運送至基隆，應該會在 5 月 9 日前送抵貴司。

寄送貨品包含二十箱的佛瑞賽特雷射水平儀，每箱約重二十五公斤。貨物將在我們的倉庫靜候提取。

您可放心，亞特拉斯物流公司將妥善處理您的訂單。我們也確信您會發現佛瑞賽特雷射水平儀擁有絕佳品質。

期盼不久的將來能再次為您服務。

謹啟，

保羅・李
佛瑞賽特精密儀器與科技

## 字彙

1. **consist** [kən`sɪst] v. 包含（+ of）

2. **await** [ə`wet] v. 等候

3. **collection** [kə`lɛkʃən] n. 收取

---

延伸學習　**出貨通知 寫作要點**

---

- **Specific cargo packaging information**
  詳細的貨物包裝資訊

- **Shipping details like forwarding agent and delivery date**
  貨運承攬商和到貨日期等的運送細節

- **Quality assurance of the consignment**
  貨物的品質保證

---

**參考答案**

1. We are pleased to confirm that the machine parts that you ordered on April 15 are packed and ready for shipping.

2. The items will be sent by Intercontinental Logistics and should arrive at your company by May 20.

3. The consignment consists of 50 cases of the machine parts, with each case weighing about 20 kilograms.

4. You can be assured that Intercontinental Logistics will take good care of your order.

# Unit 34 Payment Requests
## 請款信

## 課前寫作練習

請款信怎麼寫？
請參考主題詞彙和中譯，將下列句子翻譯成英文。

❶ 我們很高興通知您，依據訂單號碼 7534127，我們已於 7 月 5 日將掃地機器人出貨了。

 ＿＿＿＿＿＿＿＿＿＿＿＿＿＿＿＿＿＿＿＿＿

＿＿＿＿＿＿＿＿＿＿＿＿＿＿＿＿＿＿＿＿＿

主題
詞彙

- **account** [əˋkaʊnt]
  帳款
- **amount** [əˋmaʊnt]
  金額

- **balance** [ˋbæləns]
  結餘
- **due** [dju]
  應支付的

❷ 附件是供您參閱的編號 **7025057** 號發票。

(寫)

_____

_____

❸ 全額款項的到期日是 **10 月 7 日**前,所以如果您能將三萬美元的金額匯到我們的帳戶,我們將非常感謝。

(寫)

_____

_____

❹ 我們期待盡早收到您的轉帳款項,以便能完成這筆交易。

(寫)

_____

_____

參考答案請見 p. 291

- **invoice** [ˈɪnˌvɔɪs]
  發票
- **payment** [ˈpemənt]
  款項
- **purchase order (PO)**
  訂單
- **wire transfer**
  匯款

**1** **We are pleased to inform you that we have shipped the (ordered products) as per your PO# (number) on (date).**

我們很高興通知您，依據訂單號碼（編號），我們已於（日期）將（訂購產品）出貨了。

> • **We are pleased to inform you that we have shipped the aluminum alloy wheels as per your PO# CAF-0412 on May 4.**
>
> 我們很高興通知您，依據訂單號碼 CAF-0412，我們已於 5 月 4 日將鋁合金輪圈出貨了。

 • **We would like to notify you that your shipment for PO# (number) is on its way.**

我們欲通知您，您（編號）訂單的裝運貨品已出貨了。

---

**2** **Enclosed please find the invoice# (number) for your perusal.**

附件是供您參閱的（編號）發票。

> • **Enclosed please find the invoice# 6825052 for your perusal.**
>
> 附件是供您參閱的編號 6825052 號發票。

 • **Per your request, attached is invoice# (number) for your perusal.**

如您所要求，附件是供您參閱的（編號）發票。

# 3 The payment in full is due by (date), so we would appreciate it if you could remit the amount of (money) to our account.

全額款項的到期日是（日期）前，所以如果您能將（款項）的金額匯到我們的帳戶，我們將非常感謝。

- **The payment in full is due by August 3,** so we would appreciate it if you could remit the amount of **US$40,000** to our account.

  全額款項的到期日是 8 月 3 日前，所以如果您能將四萬美元的金額匯到我們的帳戶，我們將非常感謝。

 • Please remit the payment of (amount) in full to our account by (date).

請將（金額）的全額款項於（日期）前匯到我們的帳戶。

---

# 4 We are looking forward to receiving the transfer at your earliest convenience so that . . .

我們期待盡早收到您的轉帳款項，以便……

- We are looking forward to receiving the transfer at your earliest convenience so that **we can conclude this transaction.**

  我們期待盡早收到您的轉帳款項，以便完成這筆交易。

 • We would appreciate prompt payment of the amount requested so as to . . .

我們感謝您能立即支付所要求的金額，以便……

# 英文請款信

To: davis@fastautomobile.com
From: jessicachen@dcg.com
Subject: Shipment Notifications and Payment Requests
Attached: Invoice.pdf

Dear Mr. Davis,

We are pleased to inform you that we have shipped the **aluminum alloy**[1] wheels as per your PO# CAF-04121014 on June 4. Enclosed please find the invoice# 682500052 for your **perusal**.[2]

The payment in full is due by July 3, so we would appreciate it if you could remit the amount of US$40,026.20 to our account. Wire transfer details are included in the invoice.

If you have any questions, please do not hesitate to get in touch. We are looking forward to receiving the transfer at your earliest convenience so that we can **conclude**[3] this **transaction**.[4]

Best regards,

Jessica Chen
Director of Accounting
DC & G Enterprises

## 中譯

戴維斯先生您好，

我們很高興通知您，依據訂單號碼 **CAF-04121014**，我們已於 **6** 月 **4** 日將鋁合金輪圈出貨了。附件是供您參閱的編號 **682500052** 號發票。

全額款項的到期日是 **7** 月 **3** 日前，所以如果您能將 **40,026.20** 美元的金額匯到我們的帳戶，我們將非常感謝。電匯明細已含在發票裡。

如果您有任何疑問，請不吝聯絡我們。我們期待盡早收到您的轉帳款項，以便完成這筆交易。

謹致，

潔西卡・陳
會計協理
**DC & G** 企業

## 字彙

1. **aluminum alloy** [əˈlumənəm] [ˈæˌlɔɪ] *n.* 鋁合金

2. **perusal** [pəˈruzəl] *n.* 審閱

3. **conclude** [kənˈklud] *v.* 結束

4. **transaction** [trænˈzækʃən] *n.* 交易

---

### 延伸學習 「匯款」的相關用字

- **remit** [rɪˈmɪt] *v.* 匯款
  **remittance** [rɪˈmɪtṇs]

- **transfer** [trænsˈfɚ] *v.* 轉帳
  **transfer** [ˈtrænsˌfɚ]

- **wire** [waɪr] *v.* 電匯

---

**參考答案**

1. We are pleased to inform you that we have shipped the floor-cleaning robots as per your PO# 7534127 on July 5.

2. Enclosed please find the invoice# 7025057 for your perusal.

3. The payment in full is due by October 7, so we would appreciate it if you could remit the amount of US$30,000 to our account.

4. We are looking forward to receiving the transfer at your earliest convenience so that we can complete this transaction.

# Unit 35

# Late Delivery Letters
## 催貨通知

## 課前寫作練習

催貨通知怎麼寫？
請參考主題詞彙和中譯，將下列句子翻譯成英文。

**❶** 做為善意的提醒，我的來信與我訂的有機嬰兒連體衣有關，而這些品項應該很快就要出貨了。

 寫 _____

_____

主題
詞彙

- **delay** [dɪˋle]
  延誤
- **delivery** [dɪˋlɪvərɪ]
  運送；交貨
- **disappointment**
  [ˌdɪsəˋpɔɪntmənt] 失望
- **dispatch** [dɪˋspætʃ]
  發送

❷ 再次確認，我們訂購了兩百箱的鬱金香球莖，到貨日是 5 月 13 日。

(寫) _____

_____

❸ 得知貨物未準時送達，令人感到相當失望。

(寫) _____

_____

❹ 現在我們必須堅持你們在兩日內到貨，否則我們將被迫要求全額退款。

(寫) _____

_____

參考答案請見 p. 299

- **refund** [`ri͵fʌnd]
  退款
- **reminder** [rɪ`maɪndɚ]
  提醒

- **shipment** [`ʃɪpmənt]
  運輸的貨物
- **shipping address**
  運送地址

## 1 I'm writing in relation to (order) as a friendly reminder that these items should soon be due to ship.

做為善意的提醒，我的來信與（訂單）有關，而這些品項應該很快就要出貨了。

- I'm writing in relation to **my shipment for PO# 22152** as a friendly reminder that these items should soon be due to ship.

   做為善意的提醒，我的來信與訂單編號 22152 的貨物有關，而這些品項應該很快就要出貨了。

 • This is a reminder that my order of (product) should arrive on (date).

   來信提醒我訂購的（產品）應該在（日期）送到。

## 2 Just to reconfirm, we have ordered (amount of product) with a delivery date of . . .

再次確認，我們訂購了（數量的產品），到貨日是……

- Just to reconfirm, we have ordered **5,000pcs of the HITOP Water Gun** with a delivery date of **July 3**.

   再次確認，我們訂購了五千支海托普水槍，到貨日是 7 月 3 日。

 • We're expecting (amount of product) to be shipped to (address) on (date).

   我們預期（數量的產品）在（日期）送到（地址）。

# 3 It was with disappointment that I learned the shipment hadn't arrived (time).

得知貨物（時間）未送達，令人感到相當失望。

- It was with disappointment that I learned the shipment hadn't arrived **on the agreed date.**

  得知貨物未在約定日期送達，令人感到相當失望。

 • Your company promised timely delivery; however, my order has not arrived after (period of time).

貴司承諾及時交貨，然而，我的訂單在（一段時間）後仍未送達。

---

# 4 We must now insist that you provide delivery within (number) days, or we will be forced to . . .

現在我們必須堅持你們在（數）日內到貨，否則我們將被迫……

- We must now insist that you provide delivery within **three days,** or we will be forced to **cancel the order.**

  現在我們必須堅持你們在三日內到貨，否則我們將被迫取消訂單。

 • Please do everything possible to locate or reship the order so I can . . .

請盡一切可能找出或重運訂貨，我才能……

- Please let me know the status of my order so I will . . .

  請讓我知道我訂單的狀態，我才能……

# 出貨提醒

To: Wei-li Wang

From: Oscar Ramirez

Subject: PO# 20210502015

Dear Wei-li,

> 指「與……有關」，意同 with reference to。

I'm writing in relation to the order **referenced**[1] above as a friendly reminder that these items should soon be **due**[2] to ship.

Just to reconfirm, we have ordered 30,000pcs of the Supah Squirt model of water gun with a delivery date of August 30. The shipping address is 57 W. Pier St., Long Beach, CA 90802, USA. We're expecting 50pcs per box for a total of 600 boxes. Our warehouse **foreman**[3] Blake Carter will receive the order.

Please let me know when the order has been dispatched.

Best regards,

Oscar Ramirez
Head of Supply Chain and Logistics
FunnTime Toyz

中譯

偉立你好，

做為善意的提醒，我的來信與上方所提到的訂單有關，而這些品項應該很快就要出貨了。

再次確認，我們訂購了三萬支速霸噴射型號的水槍，到貨日是 8 月 30 日。運送地址是 90802 美國加州長灘市碼頭西街 57 號。我們預期每箱裝五十支，總共六百箱。我們的倉庫領班布雷克・卡特將簽收這筆訂單。

訂單出貨時請讓我知道。

謹致，

奧斯卡・雷米瑞茲
物流供應協理
歡樂時光玩具公司

## 字彙

*1.* **reference** [ˋrɛfərəns] *v.* 提及

*2.* **due** [du] *adj.* 預定的

*3.* **foreman** [ˋformən] *n.* 領班

延伸
學習　**出貨提醒 寫作要點**

- 開頭即點出賣方應準備出貨。

- 提醒賣方出貨相關事項，如品名、裝箱數量、應到貨日期、出貨地址以及收貨人姓名。

- 告知賣方出貨時應予以通知。

# 催貨信

To: Wei-li Wang
From: Oscar Ramirez
Subject: Re: PO# 20210502015

Dear Wei-li,

It was with disappointment that I learned the shipment hadn't arrived yesterday, particularly as you had actually promised delivery in advance of the agreed date.

> 指「在……之前」，意思與 ahead of 類似。

While the delay is bad enough, it is the lack of communication that is most **frustrating**.[1] If there had been an issue, you should have let us know. Not only has this greatly **inconvenienced**[2] us, we may also face losses or even legal **liabilities**[3] to our **distributors**.[4]

We must now **insist**[5] that you provide delivery within three days, or we will be forced to cancel the order and request a full refund immediately.

Sincerely,

Oscar Ramirez
Head of Supply Chain and Logistics
FunnTime Toyz

## 中譯

偉立你好，

得知貨物昨日未送達，令人感到相當失望，特別是你
曾確實承諾會在約定日期前到貨。

儘管貨物延遲已夠糟了，最讓人沮喪的還是缺乏溝通。
如果有問題，你應該要讓我們知道的。現在這個情況
不僅大大造成我們的不便，我們也可能面臨損失，甚
或是對經銷商的法律責任。

現在我們必須堅持你在三日內到貨，否則我們
將被迫取消訂單，並要求立即全額退款。

謹啟，

奧斯卡・雷米瑞茲
物流供應協理
歡樂時光玩具公司

## 字彙

1. **frustrating** [ˋfrʌˌstretɪŋ]
   *adj.* 令人沮喪的

2. **inconvenience** [ˌɪnkənˋvinjəns]
   *v.* 造成不便

3. **liability** [ˌlaɪəˋbɪləti] *n.* 責任

4. **distributor** [dɪˋstrɪbjutə]
   *n.* 經銷商

5. **insist** [ɪnˋsɪst] *v.* 堅持

---

**參考答案**

1. I'm writing in relation to my order for the organic baby bodysuits as a friendly reminder that these items should soon be due to ship.

2. Just to reconfirm, we have ordered 200 boxes of tulip bulbs with a delivery date of May 13.

3. It was with disappointment that I learned the shipment hadn't arrived on time.

4. We must now insist that you provide delivery within two days, or we will be forced to request a full refund.

# Unit 36

# Collection Letters
# 催款信

## 課前寫作練習

催款信怎麼寫？
請參考主題詞彙和中譯，將下列句子翻譯成英文。

**❶** 此信旨在通知您，您的帳款已逾期十五天了。

................................................................................................

................................................................................................

主題
詞彙

- **account** [əˋkaʊnt]
  帳款
- **balance** [ˋbæləns]
  餘額

- **invoice** [ˋɪnˌvɔɪs]
  發票
- **legal action**
  法律行動

❷ 我們希望您留意，附件中 8 月 3 日所寄的 7021 號
發票、金額為三萬美元的匯款，我們尚未收到。

⊛

❸ 關於 11 月 10 日寄給您的過期款項通知，我們尚
未收到任何回覆。

⊛

❹ 若月底前仍未收到款項，我們將不得不考慮採取法
律行動。

⊛

參考答案請見 p. 307

- **outstanding** [aut`stændɪŋ]
  未支付的
- **overdue** [ˌovə`du]
  過期的

- **payment** [`pemənt]
  付款
- **remittance** [rɪ`mɪtn̩s]
  匯款

**1** This letter is to inform you that your account is now (a period of time) overdue.

此信旨在通知您,您的帳款已逾期(一段時間)了。

- This letter is to inform you that your account is now **two months** overdue.

  此信旨在通知您,您的帳款已逾期兩個月了。

 • We have noticed that your account, which was due for payment on (date), is still outstanding.

我們注意到您應於(日期)付款的帳款仍未支付。

---

**2** We wish to draw your attention to enclosed invoice# . . . from (date) for (amount), for which remittance has not yet been received.

我們希望您留意,附件中(日期)所寄的(編號)發票、金額為(數目)的匯款,我們尚未收到。

- We wish to draw your attention to enclosed invoice# **6825** from **June 4** for **US$40,000,** for which remittance has not yet been received.

  我們希望您留意,附件中 6 月 4 日所寄的 6825 號發票、金額為四萬美元的匯款,我們尚未收到。

 • We wish to draw your attention to our invoice# . . . for (amount) which remains unpaid.

我們希望您留意,您尚未支付我們金額為(數目)的(編號)發票。

## 3 We have not received any reply to the overdue payment notice sent to you on (date).

關於（日期）寄給您的過期款項通知，我們尚未收到任何回覆。

- We have not received any reply to the overdue payment notice sent to you on **August 4.**
  關於 8 月 4 日寄給您的過期款項通知，我們尚未收到任何回覆。

- We do not appear to have received any reply to our previous request on (date) for payment still owed on this account.
  關於先前在（日期）要求支付的欠款，我們似乎尚未收到任何回覆。

- Having received no response to our correspondence of (date), we can only assume that . . .
  由於未收到（日期）所發信件的回覆，我們只能假定……

---

## 4 Should we not receive payment by (time), we will have no choice but to consider legal action.

若（特定時間）前仍未收到款項，我們將不得不考慮採取法律行動。

- Should we not receive payment by **this Friday,** we will have no choice but to consider legal action.
  若本週五前仍未收到款項，我們將不得不考慮採取法律行動。

- Unless we receive remittance of the full amount by (time), we shall be compelled to take further steps to enforce payment.
  除非我們在（特定時間）前收到全額款項，否則我們將被迫採取進一步措施以強制付款。

- We regret that we must ask for payment of the amount outstanding within (number) days.
  我們很遺憾必須要求您在（數）天內支付剩餘的款項。

# 第一次催款信

---

To: Larry Davis
From: Jessica Chen
Subject: Payment Request for Invoice# 682500052
Attached: invoice# 682500052.pdf

Dear Mr. Davis,

This letter is to inform you that your account is now one month overdue. We wish to draw your attention to enclosed invoice# 682500052 from June 4 for US$40,026.20, for which remittance has not yet been received.

If payment has not yet been made, please wire the amount in full upon receiving this letter. We would appreciate it if you could **notify**[1] us once the amount has been transferred to our account. If you have already sent your payment, please forward **confirmation**[2] of payment via e-mail as soon as possible.

Should you have any questions regarding this or any other issues, please do not hesitate to contact us. We look forward to **settling**[3] your account within the next few days.

Thank you,

Jessica Chen
Director of Accounting
DC & G Enterprises

## 中譯

戴維斯先生您好，

此信旨在通知您，您的帳款已逾期一個月了。我們希望您留意，附件中 **6** 月 **4** 日所寄的 **682500052** 號發票、金額為 **40,026.20** 美元的匯款，我們尚未收到。

若尚未付款，請在收到這封信時電匯全額款項。如果您能在款項一匯入我們的帳戶時就通知我們，我們將非常感激。若您已付款，請盡速以電子郵件寄出付款確認信。

如您對這件事或其他事項有任何問題，請不吝與我們聯絡。我們期盼在接下來的幾天內結清您的款項。

謝謝您，

潔西卡．陳
會計協理
**DC & G** 企業

## 字彙

*1.* **notify** [ˋnotəˏfaɪ] *v.* 通知

*2.* **confirmation** [ˏkɑnfɚˋmeʃən]
　　*n.*（書面的）確認

*3.* **settle** [ˋsɛtl̩] *v.* 結算

---

 **催款信 寫作要點**

---

- 開頭即點出帳款逾期的時間。

- 提供發票內容，如發票編號、日期及金額。

- 反覆告知即刻匯款的必要性。

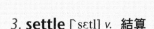

# 第二次催款信

---

To: Larry Davis
From: Jessica Chen
Subject: Second Payment Request for Invoice# 682500052

---

Dear Mr. Davis,

We have not received any reply to the overdue payment notice sent to you on July 4. We try to avoid legal action and would appreciate your **cooperation**[1] in **resolving**[2] this issue. If you haven't already made your payment, please give this matter your **utmost**[3] attention today. The deadline for remitting your payment to us in full is July 13. Should we not receive payment by the end of the week, we will have no choice but to consider legal action.

> have no choice but to + V. 指「不得不……」。

We look forward to continuing our business relationship and appreciate your **prompt**[4] reply.

Thank you,

Jessica Chen
Director of Accounting
DC & G Enterprises

中譯

戴維斯先生您好，

關於 7 月 4 日寄給您的過期款項通知，我們尚未收到任何回覆。我們試著避免法律行動，並感謝您配合解決這個問題。若您尚未付款，請在今天給予此事最大的關注。匯入全額款項的截止期限是 7 月 13 日。若本週結束前仍未收到款項，我們將不得不考慮採取法律行動。

我們期盼能繼續彼此的業務關係，也感謝您的立即回覆。

謝謝您，

潔西卡·陳
會計協理
DC & G 企業

## 字彙

1. **cooperation** [koˌapə`reʃən] *n.* 配合

2. **resolve** [rɪ`zɑlv] *v.* 解決

3. **utmost** [`ʌtˌmost] *adj.* 最大的

4. **prompt** [prɑmpt] *adj.* 立即的

參考答案

1. This letter is to inform you that your account is now 15 days overdue.

2. We wish to draw your attention to enclosed invoice# 7021 from August 3 for US$30,000, for which remittance has not yet been received.

3. We have not received any reply to the overdue payment notice sent to you on November 10.

4. Should we not receive payment by the end of the month, we will have no choice but to consider legal action.

# Customer Complaints
## 客訴信

## 課前寫作練習

客訴信怎麼寫？
請參考主題詞彙、中譯及括號內的英文提示，將下列句子翻譯成英文。

**❶** 我很不滿地投訴你們的線上音樂商店，我一直用得很不順。

---

主題詞彙

- **apology** [əˋpɑlədʒɪ]
  道歉
- **cancel** [ˋkænsəl]
  取消

- **fulfill** [fʊlˋfɪl]
  履行；完成
- **make amends** [əˋmɛndz]
  賠償；補償

❷ 這格外令人失望，因為我不斷碰到故障的問題，已對我造成極大的不便。

(寫)

❸ 最佳旋律公司（Best Tunes）的全體職員，為您無法下載所購買的音樂深表遺憾。

(寫)

❹ 做為補償，我們將提供您兩個月的免費服務，以彌補您所遇到的任何不便。

(寫)

參考答案請見 p. 315

- **measure** [ˈmɛʒɚ]
  措施
- **regret** [rɪˈgrɛt]
  遺憾
- **report** [rɪˈpɔrt]
  告知；舉報
- **tolerate** [ˈtɑləˌret]
  容忍；忍受

# 1 I am unhappy to report that ...

我很不滿地投訴……

- **I am unhappy to report that** the television I purchased from your site arrived with scratches on it.

  我很不滿地投訴我從你們網站購買的電視，送達時上面有刮痕。

 相關寫法

- It was disconcerting to find that the shoes I bought less than two weeks ago are already coming apart.

  令我感到驚訝的是，我買來不到兩星期的鞋子已開口笑了。

- I was distressed by experiencing bouts of unstable Wi-Fi connectivity.

  碰到一連串 Wi-Fi 連線不穩的問題讓我感到沮喪。

# 2 This is especially frustrating since ...

這格外令人失望，因為……

- **This is especially frustrating since** I had hoped to wear the suit to my friend's wedding this weekend.

  這格外令人失望，因為我原本希望這週末穿這件西裝去參加我朋友的婚禮。

 相關寫法

- This is totally unacceptable and not the standard of service I have come to expect from Trinity & Co.

  這令人完全無法接受，也不是我期待三合一公司會有的服務水準。

## 3 We here at (company) deeply regret (that)...

（公司）的全體職員，為 …… 深表遺憾。

> • **We here at Northeastern Brewery** deeply regret that you received a case with broken bottles.
> 東北精釀的全體職員，為您收到有破損酒瓶的裝箱產品深表遺憾。

 • We deeply regret that **the backup error that affected your account has caused you to lose information.**
對於備份錯誤影響您的帳戶，進而造成您的資料遺失，我們深表遺憾。

• We sincerely apologize for any dissatisfaction you feel with our service.
對於您因我們的服務而感到的任何不滿，我們深表歉意。

---

## 4 To make amends, we will...

做為補償，我們將……

> • **To make amends, we will** send you a replacement and give you a $25 gift card.
> 做為補償，我們將寄給您一份替換品，並提供您一張價值二十五美元的禮物卡。

 • We are initiating new procedures to preempt **this kind of mix-up from happening again in the future.**
我們正實施新流程以預防這類的疏漏於未來再次發生。

# 顧客投訴篇

To: customerservice@shopnet.com

From: stacey.chen@gmail.com

Subject: Wrong Product

To whom it may concern,

I am unhappy to report that the red **down jacket**[1] I ordered on August 25 did not arrive this morning. Instead, what I received was a black wool sweater in the wrong size. This is especially frustrating since the package took two weeks to get here despite the company promising to fulfill the order within three business days.

> 指「更糟的是」，意同 even worse。

Worse still, I had a **fairly**[2] unpleasant **interaction**[3] with customer service. When I called to **inquire**[4] about the order's **status**,[5] the representative whom I spoke with was both inhospitable and unprofessional. This is totally unacceptable and not the standard of service your site should tolerate from its employees. I would appreciate you canceling my order and refunding me for the item.

Thanks,

Stacey Chen

## 中譯

敬啟者，

我很不滿地投訴我在 **8** 月 **25** 日訂購的紅色羽絨外套，今天早上並未送達。相反地，我收到的是一件尺寸錯誤的黑色羊毛衣。這格外令人失望，因為儘管貴司承諾在三個工作天內履行訂單，卻花了兩週才把包裹送到。

更糟的是，我與客服互動的過程相當地不愉快。當我打電話詢問訂單的狀況時，與我通話的客服人員既不客氣，也不專業。這令人完全無法接受，也不是貴網站應容忍員工會有的服務水準。若你們能取消我的訂單並退還該物品的款項，我將不勝感激。

謝謝，

史黛西‧陳

## 字彙

1. **down jacket** [ˋdʒækət]
   *n.* 羽絨外套

2. **fairly** [ˋfɛrli] *adv.* 相當地

3. **interaction** [ˌɪntɚ`ækʃən] *n.* 互動

4. **inquire** [ɪnˋkwaɪr]
   *v.* 詢問（+ **about**）

5. **status** [ˋstætəs] *n.* 狀態

**客訴信 寫作要點**

- **State the problem.**
  陳述問題。

- **Provide factual details.**
  提供事實細節。

- **Stress the importance of solving the problem.**
  強調解決問題的重要性。

- **Suggest a deadline for the action.**
  建議採取行動的期限。

# 回覆客訴篇

To: stacey.chen@gmail.com
From: customerservice@shopnet.com
Subject: RE: Wrong Product

Dear Ms. Stacey Chen,

We here at Shopnet.com deeply regret the trouble you had with your order placed on August 25. Please accept our sincere apologies for any **inconvenience**[1] this may have caused you. To make amends, we will be refunding your order in full and including a voucher worth NT$1,000 that you may use for future purchases.

> 指「放心」，意同 rest easy。

> take sth to heart 指「謹記某事物」。

Additionally, rest assured that we have taken your complaints to heart. We are currently **initiating**[2] new training programs to improve our customer service experience and **preempt**[3] such **incidents**[4] from happening again in the future. We trust that these measures will be **satisfactory**[5] and hope you will give us a second chance.

Sincerely,

Trevor Baird
Customer Service Representative
Shopnet.com

## 中譯

親愛的史黛西 · 陳小姐,

購物網公司的全體職員,為您在 8 月 25 日訂購時所遇到的問題深表遺憾。對於其中可能造成您的任何不便,請接受我們誠摯的歉意。做為補償,我們會將您的訂單全額退款,並附上一張價值新台幣一千元的現金券,您日後購物可使用。

此外,請放心我們已謹記您的投訴內容。我們現正實施新的訓練課程以提升我們的客服經驗,並預防這類的事件於未來再次發生。我們深信這些措施將令人滿意,也希望您會給我們第二次的機會。

謹啟,

崔佛 · 貝爾德
客服代表
購物網公司

## 字彙

1. **inconvenience** [ˌɪnkənˈvinjəns]
   *n.* 不便

2. **initiate** [ɪˈnɪʃɪˌet] *v.* 開始實施

3. **preempt** [priˈɛmpt] *v.* 預防

4. **incident** [ˈɪnsədənt] *n.* 事件

5. **satisfactory** [ˌsætəsˈfæktəri]
   *adj.* 令人滿意的

---

**參考答案**

1. I am unhappy to report that I am continuously struggling with your online music store.

2. This is especially frustrating since I keep experiencing glitches that have already caused a great deal of inconvenience.

3. We here at Best Tunes deeply regret that you had trouble downloading the music you purchased.

4. To make amends, we will offer you two months of free service to compensate for any inconvenience you have experienced.

# Note

# Note

出版品預行編目資料

和全球做生意 必備商用英文 E-mail
= An Effective Guide to Business Writing/
希伯崙編輯團隊編著 .

-- 初版 .-- 臺北市 : 希伯崙股份有限公司 , 2021.05

面 ；　公分

ISBN 978-986-441-551-9（平裝）

1. 商業書信 2. 商業英文 3. 商業應用文 4. 電子郵件

493.6　　　　　　　　　　　　　110006162

# 《和全球做生意 必備商用英文 E-mail》讀者回函卡

謝謝您購買本書，請您填寫回函卡，提供您的寶貴建議。如果您
願意收到 LiveABC 最新的出版資訊，請留下您的 e-mail，我們
將寄送 e-DM 給您。

歡迎加入 LiveABC 互動英語粉絲團，天天互動學英
語。請上 FB 搜尋「LiveABC 互動英語」，或是掃描
QR code。

| 姓名 | | 性別 | □男 □女 |

| 出生日期 | 年 月 日 | 聯絡電話 | |

**E-mail**

□ 我願意收到 LiveABC 出版資訊的 e-DM

| 學歷 | □ 國中以下 □ 國中 □ 高中 |
| | □ 大專及大學 □ 研究所 |

| 職業 | □ 學生 □ 資訊業 □ 工 □ 商 |
| | □ 服務業 □ 軍警公教 □ 自由業及專業 |
| | □ 其他 _____ |

您以何種方式購得此書？

□ 書店　　□ 網路　　□ 其他 _____

您覺得本書的價格？

□ 偏低　　□ 合理　　□ 偏高

您對本書的評價

| | 很滿意 | 還不錯 | 普通 | 不滿意 | 很後悔 | |
| --- | --- | --- | --- | --- | --- | --- |
| 書名 | □ | □ | □ | □ | □ | |
| 封面 | □ | □ | □ | □ | □ | |
| 內容 | □ | □ | □ | □ | □ | |
| 編排 | □ | □ | □ | □ | □ | |
| 紙張 | □ | □ | □ | □ | □ | |

您希望我們製作哪些學習主題？

您對我們的建議：

縣　市

市　區
鄉　鎮

村　路
里　街

段

鄰　巷

手　號

號

樓

室

希伯崙股份有限公司客戶服務部 收